# Science in History

# J. D. Bernal

In 4 volumes

# Science in History

**Volume 3: The Natural Sciences in Our Time**

100 illustrations

The M.I.T. Press
Cambridge, Massachusetts

First published by C. A. Watts & Co. Ltd, 1954
Third edition 1965
Illustrated edition published simultaneously
by C. A. Watts & Co. Ltd and in Pelican Books 1969
Copyright © J. D. Bernal, 1965, 1969

Designed by Gerald Cinamon

Seventh printing, 1986
First MIT Press paperback edition, March 1971

ISBN 0 262 02075 0 (hardcover)
ISBN 0 262 52022 2 (paperback)

Library of Congress catalog card number: 78–136489


Printed and bound in the United States of America

# Contents

*Contents*

# Acknowledgements

This book would have been impossible to write without the help of many of my friends and of my colleagues on the staff of Birkbeck College, who have advised me and directed my attention to sources of information.

In particular I would like to thank Dr E. H. S. Burhop, Mr Emile Burns, Professor V. G. Childe, Mr Maurice Cornforth, Mr Cedric Dover, Mr R. Palme Dutt, Dr W. Ehrenberg, Professor B. Farrington, Mr J. L. Fyfe, Mr Christopher Hill, Dr S. Lilley, Mr J. R. Morris, Dr J. Needham, Dr D. R. Newth, Dr M. Ruhemann, Professor G. Thomson and Dona Torr. They have seen and commented on various chapters of the book in its earlier stages, and I have attempted to rewrite them in line with their criticisms. None, however, have seen the final form of the work and they are in no sense responsible for the statements and views I express in it.

I would like also especially to thank my secretary, Miss A. Rimel, and her assistants, Mrs J. Fergusson and Miss R. Clayton, for their help in the technical preparation of the book – a considerable task, as it was almost completely rewritten some six times – and its index.

My thanks are also due to the librarians and their staffs at The Royal Society, The Royal College of Physicians, The University of London, Birkbeck College, The School of Oriental and African Studies, and the Director and Staff of the Science Museum, London.

Finally, I would like to record my gratitude to my assistant, Mr Francis Aprahamian, who has been indefatigable in searching for and collecting the books, quotations and other material for the work and in correcting manuscripts and proofs. Without his help I could never have attempted a book on this scale.

J. D. B.

*1954*

# Acknowledgements to the Illustrated Edition

For the preparation of this special illustrated edition of *Science in History*, I must thank, first of all, Colin Ronan, who chose the illustrations and wrote the captions.

I should also like to thank Anne Murray, who has been responsible for correlating all the modifications involved in producing a four-volume version and for correcting the proofs.

Finally, I thank my personal assistant, Francis Aprahamian, who advised the publishers at all stages of the production of this edition.

J. D. B.
*1968*

# Note

In the first edition of this book, I avoided the use of footnotes. A few notes have been added to subsequent editions and are marked with an asterisk (*) or a dagger (†) (if there is more than one footnote on a page). The notes have been collected together at the end of each volume and are referred to by their page numbers.

The reference numbers in the text relate to the bibliography, which is also to be found at the end of each volume. The bibliography has eight parts that correspond to the eight parts of the book. Volume 1 contains Parts 1–3; Volume 2 contains Parts 4 and 5; Volume 3 contains Part 6; Volume 4 contains Parts 7 and 8.

Part 1 of the bibliography is divided into three sections. The first contains books that cover the whole work, including general histories of science. The second section contains histories of particular sciences and the books relevant to Part 1. The third section lists periodicals to which reference has been made throughout the book.

Parts 2, 3, 4 and 5 of the bibliography are each divided into two sections. The first section in each case contains the more important books relevant to the part, and the second the remainder of the books.

In Part 6 of the bibliography, the first section contains books covering the introduction and Chapter 10, the physical sciences; and the second section, Chapter 11, the biological sciences.

Part 7 of the bibliography contains books covering the introduction and Chapters 12 and 13, the social sciences.

Part 8 of the bibliography contains books covering Chapter 14, the conclusions.

The system of reference is as follows: the first number refers to the part of the bibliography; the second to the number of the book in that part; and the third, when given, to the page in the book referred to. Thus 2.5.56 refers to page 56 of the item numbered 5 in the bibliography for Part 2, i.e. Farrington's *Science in Antiquity*.

# Science in Our Time

# Introduction to Part 6

**The Twentieth-Century Background:**
**The Revolutions in Science and Society**

As we reach our own times history blends into remembered experience. Here we are close to the events, are watching struggles still unresolved, with their protagonists still alive and active. All this makes it especially difficult to comprehend what is happening, to analyse and judge the significance of the movements of science and society. Yet despite this the effort must be made, for while it may suit historians in general to avoid dealing with recent periods until time allows disinterested appraisal, that is doubly impossible here. A book that sets out, as this one does, to show the connexions between science and social forces can be useful only if it can show how those relations, as we find them here and now, have arisen out of their previous history. No gap can be allowed between present and past. But to omit the story of science in the twentieth century would exclude the most important part of the whole argument, for it is in this twentieth century that science has come for the first time into its own. Far more scientific work has been done in the last sixty years than in the whole of previous history. And this is no mere quantitative growth; at the same time there has been greater advance in the knowledge of the fundamental nature of matter, animate and inanimate, than in any comparable period in the past. We may reasonably speak of a second *scientific revolution* in the twentieth century.* Further, and this touches more closely the purposes of this book, for the first time in history science and scientists have been involved directly and overtly in the major economic, industrial, and military developments of their time.

The problem is no longer, as was the case in the earlier chapters of this book, to demonstrate how science has affected the course of history. The effects of science in the past were real enough, but they had to be sought for. The danger had been that science would be thought of as an appendage – interesting, brilliant, but remote from the main stream of history. Now, half-way through the twentieth century, the danger is the opposite one, that of giving science too much credit for good or ill in the tremendous and disturbing changes, the wars and revolutions, that this century has already witnessed.

It is no accident that the revolutions in science and society should occur together, but it would be too simple a view to make either one the consequence of the other.* The interactions have been far more subtle and reciprocal, and their disentangling will be the main task of the remaining chapters of this book.

What need to be sought out, at every major turn of events, are the social and economic forces that have helped to determine the general directions and speeds of scientific advance, and conversely, the points at which scientific discoveries have come to modify profoundly the course of economic and even political events.

A TIME OF TRANSITION

The events, terrible, rapid, confusing as they have been, are not without a general pattern. We are living in an age of transition from one kind of society to another, in the middle of conflicts still unresolved. The division of the world which first appeared in 1917 is an index of the sharpness of the contrast between the old and new forms, but it only brought out into the open conflicts already latent in the apparently uniform society of the nineteenth century. However differently people may feel as to the explanation and the outcome of the struggle no one can deny its existence. The whole system of *capitalism*, first established 300 years ago, is now being challenged by another, *socialist*, system which has arisen out of the inner conflicts of capitalism itself.

For most of the twentieth century, however, it is not the open challenge represented by the existence and growth of the Soviet Union that has been the main determining factor in world history. It is rather the continued working of forces from an earlier time. Two of the decisive events of the century, the First World War and the great slump of 1930, were the products of political and economic difficulties wholly inside capitalism, and so were both the preparations and the early stages of the Second World War. The evolution of capitalism went on for the whole period and it is still the dominant economy over a vast, if diminishing, portion of the world.

The evolution of the socialist part of the world, first in Russia alone and now in China and in many other countries, has necessarily been of a different kind. Partly on account of the initial poverty of the countries and partly because of the hard struggles that have had to be waged to build up a radically new economy, in the face of continuous interference from external enemies, it is only in very recent years that the socialist countries have begun to claim a leading part in world economy, technology, and science.

Nevertheless, in spite of this lag, the importance of the developments in the socialist countries is far greater than their mere scale would indicate. They represent a new kind of way of employing natural and human resources which is impressing the workers of capitalist countries and even more the peoples of the under-developed countries. These have won some measure of political freedom and are now demanding effective economic liberation, an additional and powerful element in the transition from capitalism.

## MONOPOLY AND IMPERIALISM

In the capitalist world the major feature of the twentieth century has been the rapid growth to complete dominance of large combines, trusts or cartels, partly commercial, partly industrial. Even their names are familiar all over the world – Du Pont, General Motors, Krupp, Schneider Creusot, Imperial Chemical, I.G. Farben, etc., not to mention the nominally dispersed empire of Standard Oil or the wide range of Morgan interests. The tendency to monopoly, already evident in the late nineteenth century, has in the first place an economic source. Trusts, exercising partial or complete monopoly, had great advantages over small competitive firms in securing profits, no longer at the mercy of market fluctuations, and in tiding over bad times. They were also favoured by technical factors, such as the development of the internal-combustion engine creating the motor industry and providing in turn vast markets for a new oil industry. The technical innovations themselves, such as mass-production, raised the amount of capital necessary for manufacture on a scale large enough to be profitable to a level only monopoly firms could reach. Finally, science itself has helped the formation of monopolies through the same requirement of large capital outlay. The industries mainly or entirely founded on science, such as the chemical and electrical industries, were monopolistic from the start. As a consequence, as we shall see (pp. 1253 ff.), some eighty per cent of industrial science is carried out in the research departments of monopoly firms.[6.66; 6.67]

The very existence of trusts and cartels provides protection for prices well above the competitive level. This, combined with the reduction in costs obtained by large-scale production, making fuller use of engineering and scientific research, has helped the monopolies to secure ever greater profits. They have consequently been able to increase in range through mergers and new ventures. Their network of control, only part of which is ever made public, puts them in an apparently unassailable economic position. As productive enterprises they undoubtedly mark an improvement on the small traditional, rule-of-thumb firms that they broke or absorbed. Nevertheless experience has shown that they are no better able to escape the nemesis of all production for profit. The greater the efficiency of exploitation of labour they have achieved the more difficult it is to find consumers from among those same workers for the goods they produce. It has been the need for new markets and for the protection of those already acquired that has led monopoly interests virtually to take over the functions of government to further their own purposes (p. 1134).

From 1880 onwards government policy, particularly foreign and colonial policy, has largely been dictated by the urge to secure greater and greater shares of the world markets for the products of monopoly enterprises, especially in the exports of such capital goods as steel and machinery. This is the pattern of *imperialism* – once proudly flaunted, now a reproach which needs to be explained away – which in one form or another, under the Union Jack or the Stars and Stripes, remains the dominant form of capitalism.

Despite arrangements arrived at from time to time to share out the markets of the world among the monopolies of different countries, these could not be lasting and rivalries tended to increase. Whenever the allocation of markets seemed no longer to correspond to the real strength of the powers, the only means of changing this was military force. Hence the many wars, small and large, which have plagued the world these last seventy years. War and war preparations have also themselves been an essential outlet for the products of the most powerful monopoly firms in the steel and chemical industries. They have provided unlimited orders and no excessive scrutiny of prices. The vast and ever growing military expenditure of the Cold War period has to some extent eased the problem of disposing of the surplus product of mechanical production. Indeed, war preparation has become a necessary part of the economies of all the large capitalist countries, with the additional advantage of imposing added burdens on the economies of socialist countries trying to produce for use rather than profit. Disarmament is feared and is being continually put off by the Western powers, as much for economic as for political reasons.

These background conditions can help to explain the differences in the rate of advance and in the kind of use of science in the world of the latter twentieth century. The intrinsic common basis of all science, founded in the character of the physical and biological world, leads naturally to a convergence between the scientific and technical pattern all over the world. The divergences are primarily of political and economic character. For the purposes of this book the world can be considered as divided into three sectors. The first is that of the capitalist world, including Western Europe, North America and their outposts in such countries as Australia and South Africa and the one Asian country, Japan, that became industrialized while remaining independent in the imperialist era. The second comprises the socialist countries, at first only the Soviet Union, then other countries of Eastern Europe, then the great People's Republic of China with its resurgent ancient civilization, and more recently the liberated countries of Northern Korea and Vietnam. The third category includes the rest of the world, not yet socialist and counted politically as neutralist, though economically effectively part of the capitalist 'free world'. It is far more heterogeneous, including Latin America, with the legacy of the renaissance culture of colonial Spain and Portugal, India and Arabic Asia and North Africa, with their old Hindu and Islamic cultures, and, finally, the only recently and partially liberated countries of Africa and Asia. In each of these sectors science has different achievements and prospects which will have to be considered in their place after the general programme of advance of science has been considered.

In the meanwhile these brief paragraphs may serve as an introduction to the political and economic background of our times. A more critical appreciation is reserved till after the discussion of the social sciences in Chapter 13 (pp. 1249 ff.).

## THE PLACE OF SCIENCE AND TECHNOLOGY
## IN THE ERA OF MONOPOLY

The tightening of the links between monopoly, imperialism, and war has had the effect of bringing capitalist governments, whose primary responsibilities and greatest expenditure are on armaments, directly into the development of new weapons to be manufactured by the big monopoly firms. These weapons – jet planes, guided missiles, ballistic rockets, atom and hydrogen bombs – are becoming increasingly scientific, not only in original invention, but in constant subsequent improvement. They consequently involve governments in scientific research and development, growing at an enormously rapid rate. Military research expenditure already vastly overshadows not only that on pure science, but also even that on industrial research (p. 846 f.).

The effect on science of the nationalization of industries has been a very minor matter in comparison with its military commitments. This is due to the fact that as these industries were unprofitable under private enterprise very little research was done on them, and that now, under nationalization, it is of a very low priority. On the other hand the virtual taking over of the finance of universities by governments in Britain, and even in the citadel of free enterprise, the United States under pressure of defence research contracts, has made an enormous difference to the status of research. Although for the time being the control exercised over research, at least in Britain, is very indirect, it does in fact mean that the general direction of fundamental research has now passed into governmental hands.

While these processes of concentration of power were taking place the independent competitive capitalists who dominated the nineteenth-century economies were rapidly being submerged. It is not that there is no room for the small man. Actually the ancillary requirements of modern large-scale industry offer opportunities for innumerable sub-contractors and component suppliers. It is rather that their relative importance has shrunk; they depend on the big firms; they have become clients and have lost their independence. The same loss of status has befallen the inventors and amateur scientists who played such a large part in advancing science since the seventeenth century. From now on scientists and technologists alike, together with most of the doctors, have ceased to be professional men in the old sense, exercising their skill for fees or working on their own account, and have become employees or executives of government departments or large firms.

This change, which has come about, at first gradually, then very rapidly during and after the Second World War, is bound to have a profound influence on the attitude of scientists not only as individuals but also in relation to their work. It creates a deep conflict between their immediate dependence on their source of livelihood and their responsibility for the safeguarding, advancement, and use of science, a problem to which we shall return later (pp. 1260 ff., 1282 f.).

SCIENCE IN SOCIALIST ECONOMIES

THE SOVIET UNION: So far I have discussed only the economic trends that have affected science in capitalist countries. Its development in the Soviet Union and in other countries that have taken decisive steps towards socialism has been very different. There, where all major industries have been taken over by the State, where there is neither monopoly nor competition, there is a deliberate and conscious drive to develop and use science to the full. This has been achieved not by subordinating science to the industrial and agricultural organizations that have there taken the place of private firms, but rather by using the old academies (p. 1265) and turning them from the honorific societies they had become into active centres of research and higher teaching. It is the scientists, grouped in the academy and its institutes, and increasingly now in the universities, who plan this work, with the object of securing at the same time the most fruitful intrinsic growth of science and the maximum of help that can be given to the full utilization of natural and human resources. This will be referred to again in connexion with various aspects of scientific work (pp. 829 ff., 1170 ff.).[6.16; 6.52; 6.104] (cf. p. 705)

THE PEOPLE'S DEMOCRACIES: The pattern of scientific teaching and research in other countries of Eastern Europe – Poland, Hungary, Rumania, Bulgaria and the German Democratic Republic – has been reconstructed along the general lines, though with considerable national differences, of that in the Soviet Union. The main feature is the revival of the Academies and the linking of scientific research with the development of industry and agriculture. All these countries have to a greater or lesser degree a long established tradition of science and a correspondingly developed industry. The profession of science was, however, largely limited to a small hereditary intelligentsia from which workers and peasants were virtually excluded. The great expansion of higher education has brought them in and we may confidently expect a great contribution to world science from them.

THE PEOPLE'S REPUBLIC OF CHINA: The situation in China after the Liberation in 1949 was different in many respects from that of the Soviet Union and the countries of Eastern Europe. This is not only because it is an enormous country, among the largest and certainly the most populous in the world, and one with a very old and absolutely continuous tradition of scholarship, but also because modern science only entered China in the wake of imperial exploitation of the country, which had continued in an aggravated form of almost continuous warfare and oppression for the last half of the nineteenth and the first half of the twentieth century.

As a result it was also a country with a low standard of living and, in fact, a typical under-developed country until 1949. The changes that have taken place since Liberation, however, have been completely different from and far greater than any that have occurred anywhere else in the world and the miracle of reconstruction has been achieved on the basis of the full development of human as well as natural resources, particularly by the enormous drive for education (pp. 1198 f.).

An attempt is being made to build up in a few decades a body of scientific and technical workers adequate for a modern industrial State. Because the country cannot wait even so long for the utilization of its natural resources and the building of its heavy industry, special efforts are being made to train geologists and metallurgists on a scale already much greater than obtains in capitalist countries.* At the same time research is being actively fostered under the general control of a revived Academia Sinica, which, as in the Soviet Union, combines the development of fundamental science and a service to the economic needs of the country. First-class work is already being done, and it will not be long before China moves into the forefront of world science.

SCIENCE IN THE UNDER-DEVELOPED WORLD

The great events of the last two decades which resulted in the liberation, at least politically, of some thousand million people in Asia and Africa, combined with the reawakening of movements towards economic independence in the countries of the former Spanish and Portuguese empires in Latin America, is again of a different character from that of either the capitalist or socialist parts of the world. Only in Cuba in 1959 has a Marxist Socialist state with its full accompaniment of land reform, industrialization, education, with the elimination of illiteracy, and scientific research, been achieved in the new world. For historical reasons the economic situation in all the other countries is still largely subsidiary to those of capitalist economies.

Moreover, the basic economic function of these countries was to provide raw materials for the industrial countries of Europe and North America and not to develop their own industry. Consequently their own science, the scientific pattern before liberation and immediately following it was of a peculiar and restricted kind, largely limited to medicine, to ensure a healthy working force, to cash crop production and the processing of raw materials. Though minerals represented a very large part of the products of these countries, research in geology and mining was mostly carried out in the industrial countries themselves. Development of all-round science had to wait for liberation and for the corresponding building up of industries in the countries themselves, supplying the necessary goods for trade and consumption, a process which has not yet proceeded very far even in such a country as India, which had the longest history under British rule.

Everywhere in these countries there is a move in this direction which has been helped by the setting up (in many of them) of governments of a more or less socialist character, at least to the extent of wishing economic as well as political independence. Some of the worst features are the feudal land-holdings, effectively inimical to the development of science. Of these countries, India and the Arab states have had the longest experience of modern science, which they have successfully grafted on to their culture, most of it much older than any to be found in Europe. Though here and there the appearance of brilliant individual scientists has brought them into the front rank for the production of science, there is still a long way to go, especially in those countries where illiteracy looms large and drastically limits the possible field for the recruiting of scientists.

INTERNATIONAL CO-OPERATION

To a certain extent international organizations such as UNESCO have been helpful in bringing the science in all these countries in contact with that of the more rapidly advancing industrial countries. Yet it is to be feared at the moment that whatever the speed of advance, it is slower than that of the advancing countries and the gap between them is on the whole widening rather than narrowing. Much greater effort will have to be made before science in all these countries enters into the general field of scientific effort in the world.

The beginning of such an effort was seen in the United Nations Conference on the Application of Science and Technology for the Benefit of the Less Developed Areas (UNCSAT) held in Geneva in February 1963.[6.136] This conference however, on account of its sponsorship, was defective in that it avoided any reference to the economic and political difficulties: feudal landowning, cash crop production and foreign mineral and services ownership, which prevent science being effectively used in these areas. In particular, the absence of the People's Republic of China deprived the participants of learning of the most extensive and rapid use of science in building up the economy, welfare and education of the largest country of the world. This defect had been avoided in a smaller conference held in Warsaw in 1959 on the same general theme by the World Federation of Scientific Workers.[6.13]

INTERACTIONS OF INDUSTRY AND SCIENCE

Modern industry is permeated by science, and in certain lines, such as electricity and chemistry, it is largely a scientific creation. It is therefore no longer relevant, as it was in earlier times, to describe the specific characters of industry and to follow with their influence on scientific thought. The degree of inter-penetration is already too great. It is only worth attempting to bring out the general character of the influence of technology on science and to illustrate particular interactions as they arise in later chapters.

The technical developments of the twentieth century already indicate that we are in the presence of a second or rather third major industrial revolution (pp. 854 f.). The comparison may, however, obscure the fact that it is a revolution of a new kind, one in which planned scientific research is taking the place more and more of individual mechanical ingenuity. Further, while the great Industrial Revolution was concerned largely with the production and transference of force, relieving men, in principle, from hard muscular work, the twentieth-century revolution is largely in the substitution of the machine or electronic device for the skill of the worker, and should relieve him from the burden of monotonous clerical or machine-minding tasks.

Although the first steps to such a revolution have been taken in the development of automatic and servo-mechanisms, this has been only a recent achievement. The earlier features of twentieth-century industry were more in the line of expanding and extending into new fields the devices of the nineteenth century. The impulse to the direction of twentieth-century technology is provided by the special profitability of the mass media of transport, communication, and entertainment.

In transport the motor-car, the tractor, and the aeroplane were made possible in the first place by the internal-combustion engine – itself a nineteenth-century development. These substituted for the rigid and limited facilities of the railway the flexibility and range of millions of small units that could go everywhere and do anything.

To make these for a great, new, low-price market meant the rapid spread of mass-production methods. The motor-car and the motor industry called in turn for vast extensions in the production of petrol, rubber, sheet steel, and plastics, which promptly found a multitude of other uses. A new engineering and light industry grew up in which centrally generated electricity took the place of the stationary steam-engine, and this, together with the entry of electricity into the home, created a new heavy electrical industry. Less important economically, but more noticeable and carrying a larger contribution from science, were the new electrical communication industries of radio and television and the exploitation of photography in the cheap Press and the cinema.

This catalogue unfortunately cannot be exhausted with the peaceful uses of technology. The aeroplane has had, almost from the outset, a primarily military objective, from which civil aviation can take some pickings. War is also responsible for the multiple refinements of electronics in telecommunication and radar and for the new lethal interest in atomic energy.

Underlying the mechanical and electrical devices, though far less conspicuous, has been the rapid growth of a new all-pervasive scientific chemical industry producing everything from fertilizers to detergents, from nylon to antibiotic drugs. It was ready to turn out explosives and gas for war, and now it has become a mainstay of atomic and power production.

POWER AND CONTROL

The multifarious productions of science among which we live more and more of our lives depend largely on the use of two very general and extremely important new technological principles. The first is the availability of *power* in adequate quantities just where it is needed, whether in beating an egg in the kitchen, turning a twenty-ton casting in a factory, or cutting down a tree in the distant forests. This service which electric grids and the ubiquitous petrol engine provide between them is one reason for the more than five-fold increase in productivity per man hour that has been achieved in the last fifty years in the United States.

The second principle, likely to be even more important in the future, is that of precise and increasingly *automatic* control of all industrial operations, whether mechanical or chemical. Many chemical plants have already become fully automatic, with electronic devices keeping all the variables in control. In engineering, fabrication and assembly lines are well set on the same path. Between them, these two principles imply ever-increasing strength and skill which science makes available to industrial processes as a whole, thus supplementing and extending without limit the range of the craftsman's arm and brain. Of the two, the first is but an extension of the mechanical power of the Industrial Revolution. The second is something radically new, an extension of human senses, nerves, and brain by electrical means, and, through the unlimited range of the combinations it can offer, is bound to have unpredictably greater material and social effects (pp. 855 f.).

These developments are now beginning to make themselves felt. Atomic power and automation have arrived. In earlier stages the major changes have been due to the increased size and concentration of plant. It is this that has made possible the multiplication of industrial research laboratories which range from mere testing shops to almost university rank. What occurred quite exceptionally in the late nineteenth century (p. 564) is now the rule. Science has now won a definite place in industry. This, combined with the growth of similar laboratories in the government service, means that the interaction between science and the productive processes generally has now become much closer and more important. It has indeed become something radically different in the twentieth century from what it was in earlier ages: it is on a larger scale, it is much more rapid, and it is becoming a fully conscious interaction.

## THE SCALE OF SCIENTIFIC ADVANCE

The scale of scientific effort. has in the twentieth century increased almost out of recognition. In 1896 there were perhaps in the world some 50,000 people who between them carried on the whole tradition of science, not more than 15,000 of whom were responsible for advancing knowledge by research. Sixty-six years later there were at least a million active research workers, and the total number of scientific workers in industry, government, and education is almost impossible to assess accurately, but must approach two million people. The expenditure on science has increased in far greater proportion, from less than half a million to nearly £10,000 million, an increase of 2,000 times allowing for the change in the value of money. This implies an average rate of growth of ten per cent per annum.[6.115] The rate of increase in the last few years has been much greater, up to twenty-five per cent (p. 847). Such rates of growth are far greater than those of any other element of society, greater even than that of military expenditure. Science, however, is still a long way behind, for though some ninety per cent of scientific expenditure is for war research and development, this is only twelve per cent of military expenditure.[6.114; 6.125]

Such a rate of growth betokens more than a mere change in size, it is in itself an index of a profound change in the character of science and in its relations to society. Of that change there are ample indications from inside science and in the ever-growing dependence of industry and government on science. That dependence has become completely reciprocal. Not only has the total cost of science increased out of measure, but so also has that of its separate components. Even apart from the multi-million-dollar machines, now indispensable in many fields of physics research, ordinary laboratory costs are beyond the purse of all but the wealthiest individuals or even of most teaching institutions, forcing dependence on big business or government.

Another significant feature of the transformation is the change in geographical location. In 1896 practically the whole of world science was concentrated in Germany, Britain, and France, the remaining centres of science in Europe and America being in effect subsidiary local branches of the science of those countries, and there being comparatively little science in Asia and Africa. By 1954, while the science of the old centres had grown considerably but unevenly, that growth was quite eclipsed by the enormous development of science in the United States and the Soviet Union. Japan and India have been making substantial contributions to the advancement of world science since the beginning of the century, but along essentially Western European lines. The liberation of China has added a new dimension to the building up of science on a popular basis on a far larger scale and directly linked to the needs of a rapidly developing economy. This pattern is already spreading to other Asian countries such as Korea, Vietnam and Indonesia. In the measure that colonialism is forced to retreat, the need for higher education and research is expressing itself in different ways in former dependent countries in Africa and meets with a renewal of scientific interests in the predominantly classical culture of Latin America, as the example of Cuba is now showing.

A world science is indeed in process of formation, and one consciously linked from the very start with the expansion of industrial and agricultural production. It should be noticed also that though the philosophy of science differs widely between socialist and capitalist countries, as also do the major uses to which science is put, both types of system have come to need science more and more urgently.*

THE RAPIDITY OF APPLICATION IN SCIENCE

The third characteristic of science in the twentieth century is the far more immediate and rapid application of scientific discoveries. Although it still remains true that the science on which the bulk of twentieth-century technique is based is nineteenth-century science – in power production, in electricity, and in chemistry – inventions depending entirely on more recent discoveries have made their impact in minor but striking roles. Radar and television, plastics and artificial fibres, synthetic vitamins, hormones, and antibiotics, all are but the first samples of what will come from the great scientific revolution of the twentieth century, and so also, if we are not careful, will the large-scale use of atomic and hydrogen bombs, radioactive and bacterial poisons. These are just examples of a principle more important than all of them put together, that of the universal possibility of using natural science, immediately or with a delay of a few months or years at most, in the formulation and solution of any problem in practical life. What happened almost by chance in the nineteenth century, or was brought about by the genius and force of character of a solitary inventor like Bessemer or a public-minded scientist like Pasteur, is now a recognized and almost routine way of tackling industrial, agricultural, or health problems.

Indeed, we have reached the stage at which it is foolish and self-defeating to leave such problems to the old stand-by of chance or rule of thumb. Research and development have become recognized disciplines embodied in rapidly growing institutions. Science has now entered industry in an intimate and operational way, and in doing so has been both enlarged and transformed. Nor has the development stopped there. The increasing scale of scientific application and the urgency that war and war preparations impresses on it have involved science ever more closely with governments, while in the newly established socialist countries science is necessarily invoked from their very inception in every constructive scheme. It is from this experience of science that has grown a new consciousness of its power as an agent in social transformation. A modern community has come to depend on science for its very existence. We are beginning to see in this country the realization of the hopes of the men of the seventeenth century like Descartes, when he declared that through science we could 'become the masters and possessors of Nature' (p. 447).

Today we are participating in the culminating point of the revolution which such men as these started 400 years ago. It is one which is comparable in importance with that which ushered in the first human societies; it is even greater, because of the unlimited further prospects it offers, than that which followed the invention of agriculture. It is now apparent that man is about to reach a state in which he can control his material environment through the conscious use of science.[6.19] He can secure himself against want, abolish tedious toil, and by rapid stages reduce the misery of disease. How far this will be done is now seen to depend square on man's ability to adapt his social forms to provide the co-operation that is needed to secure these aims and to overthrow the interests that stand in the way. Thus the science of human society and of its law of transformation comes to occupy the central place in the determination of the future.

The power of science to affect the life of man for good or evil is no longer seriously in doubt. The problem now is rather that of finding the means of directing science to constructive and not destructive ends.

This, however, is a problem much greater than any in the particular sciences which we are considering. We will return to it at the end of Chapter 14, when considerations from the physical, biological, and social sciences can all be brought to bear on it. Here it is sufficient to consider the more immediate and practical question of the most rapid utilization of science or of the means of closing the gap between scientific ideas and their practical utilization. This gap, which was formidable in the nineteenth century, existed for reasons that were primarily economical and not technical (pp. 612 f.). It took the abnormal conditions of the two great world wars to prove in practice that it could be narrowed, and to show the way this could be done even in peace.

EFFECTS OF WAR ON THE ADVANCE OF SCIENCE

The First World War, which fostered the development of the bombing aeroplane, the tank, and poison gas, gave some foretaste – an exceedingly bitter one – of what science could do in war. By bringing scientists and practical men directly together with the incentive of military requirements and relatively no restrictions on funds, it forced the recognition that there was no need to wait for years before putting an idea step by step through experiments and trials into full production. This lesson was no sooner learned than it was largely forgotten, as witness the slow pace of the development of such obvious winners as the jet engine (p. 812) and television (p. 780) between the wars. The Second World War was needed before the lesson could be accepted and acted on. The first spectacular proof of this was the production of the atom bomb – from the scientific discovery of atomic fission as a hardly detectable effect in 1938 to a death-dealing horror in 1945, with the expenditure of more money than science had used in the whole course of human history up to that time. The advent of the Cold War gave a further impulse to science in the service of destruction that was to dwarf all these earlier efforts. It has led to an unprecedented rate of growth of science with an ever-growing proportion devoted to weapon research. In the following chapters we shall be tracing the effects of this distortion on science and scientists (p. 759).

SCIENCE AND PLANNING

War produced the most outstanding example of the conscious use of science in the twentieth century. In all fields of industry and agriculture this new integrated approach began to be used. Indeed, it was from the outset the policy of the new socialist society brought into being by the Revolution of 1917. Industry, agriculture, medicine, and even science itself began to be planned instead of being left to the chance of economic forces. For all their overt disapproval, industries and governments in capitalist countries began to copy the Soviet Union in its tendency to plan. In the light of experience of successes and failures it began to be seen that the applications of science did not just come of themselves, but that human needs had first to be discovered and that then deliberate and planned scientific effort was needed to find the means to satisfy them. This dawning consciousness of the function of science was one of the most characteristic features of the twentieth-century social revolution. It corresponded with an equally far-reaching, but as yet also incomplete, revolution inside science itself.

The great and terrible events of the time – crises, wars, and revolution – whatever they might import to the main ends to which science and technology were used, were, as we all know, quite compatible with a great new efflorescence of science. The stream of new discoveries and inventions, the depth and range of the new scientific theories are, however, for all their novelty, but continuations of internal movements of scientific experiment and thought that have been progressing ever since the Renaissance. The inner nature of the advance of science in our time can be accounted for by reasons drawn from the internal history of science, though even here the influence of external factors has often been great. Nevertheless, the unprecedented *scale* and *speed* of the whole movement are linked directly to technical and economic factors. So also are the general *strategy* of advance and the relative effort devoted to the different fields of science (pp. 1288 f.).

SCIENCE PAYS ITS WAY

The major and decisive fact is that, starting in the nineties, and with a momentum rapidly increasing in the First and Second World Wars, science began to pay its way. It became, fully consciously and immediately, what it had long been unconsciously and incidentally – an essential part of production. It was something worth investing in, directly by setting up research laboratories or indirectly by subsidizing universities where the workers for these laboratories could be trained and where basic research, of use to all, could be carried out.

In the course of fifty years a complete transformation of the position of science in society was effected, in which three stages are already distinguishable. At the beginning of the period, in the nineties, we are still in the era of *private* science, that of the small laboratory of the professor or the back room of the inventor. The next stage, first evident in the twenties and thirties of the new century, is the era of *industrial* science, that of the research laboratory, spending a few tens of thousands of pounds, and of the correspondingly expanded university department and the now subsidized research institute. The third stage, appearing first in the Soviet Union but becoming universal in the Second World War, is that of *governmental* science, where the expenses of research and development run into hundreds of millions of pounds and establishments as large as towns are needed to house the men and equipment needed for it. For this only the State can find the money, though it may call on the assistance of monopoly firms, themselves almost States in their own right, to spend it for them in the form of development contracts.

With each increase in the scale goes an increase in the scope of the application of science. In the first stage it is for detailed improvements and small devices. In the second it is for whole new scientific industries – for radio or fine drugs. In the third stage science reaches the greatest enterprises – the war production that has been made the focus of State capitalist enterprise; or the great constructive and Nature-transforming projects of socialism.

SCIENCE AND EVERYDAY LIFE

With this expansion of scientific effort has gone a two-pronged extension of science into the processes of industry and into the apparatus of daily life. Science is at the same time becoming more useful and more familiar. Every phase of industry and agriculture is now permeated with science, and more and more consciously so. Scientific instruments are used, and scientific concepts are replacing immemorial traditions on the bench and in the fields.

The same tendency now spreads to the home. Not only are the most elaborate scientific devices, such as television receivers, becoming familiar, but in the daily routine of cooking and washing, in the care of children, in the preservation of health and beauty, the products and the ideas of science are making their way. Not all the deceits and fables of advertising are sufficient to prevent the spread of a new serious and exciting interest in science. Indeed this interest produces in turn a practical impetus to science. The popular market for scientific gadgets is becoming a major source of profit, and this helps research; while the popular interest in science itself has brought into being a new profession of *scientific journalism* and an avidly read *science fiction*.

## THE STRATEGY OF SCIENTIFIC ADVANCE

These general considerations, while they go some way to explain the rapid increase in the scale and tempo of science during the century, need to be examined more closely before any account can be given of the particular directions taken in the sectors of scientific advance. Only in certain cases, and these not the most important scientifically, have economic needs had an immediate effect on the advance of specific sciences. An example is furnished by the dependence of the study of atmospheric electricity on the development of wireless communication and the subsequent application of the principles of reflection in radar (pp. 777 f.). More commonly the impulse has been given by the inner developments of the sciences, and these have blossomed out wherever these developments have found extensive and profitable applications in peace or war. Examples are the wide search for antibiotics following the isolation of penicillin (pp. 925 f.), and for the atom bomb following the discovery of nuclear fission (p. 759). This type of relation between science and society, in earlier times, has already been described. What marks out the twentieth century is both the enormous scale of the industrial activities based on science and the rapidity of the interactions between scientific and technical advance. What some of those inter-actions were will be shown in outline in the succeeding chapters.

## THE SCIENTISTS' REACTION TO HISTORICAL EVENTS

The effects of internal developments in science and those of technical and economic factors are, however, not even together enough to account for the character and the spirit of the twentieth-century advance of science. Much weight must be given as well to the influence on the minds of the scientists themselves of the great events among which they lived, and the material and moral problems which their increasingly important participation and responsibility brought to them personally. Such influences were general and not specific, and it is impossible to attribute to them particular advances in science. They did, however, tend to draw workers to or repel them from such disputable fields as nuclear physics and microbiology in the measure that these became identified with atom bombs or bacterial warfare.

The most prevalent reaction of the scientists was to bar uncomfort-able facts from their consciences, but this process itself meant turning their scientific interests in a more abstract or, as they would have put it, in a more purely scientific direction. The stubborn insistence of some scientists on the purity and freedom of science is itself an indication of an uneasy conscience as to the social consequences of their work and as to the effects of social changes on the future of science itself (pp. 1282 ff.). On the other hand, a small but increasing number saw and welcomed the break-up of the old order, and understood how science itself could be a liberating force, both in its indirect effects through transforming industry and directly by widening all men's minds and giving them a greater possibility of realizing their capacities. As a result of these diverging tendencies science was torn with conflict, but that in itself may actually have assisted its progress, because science has always grown on criticism, and especially in the twentieth century no theory or dogma was safe. Internally science was being attacked as a result of its own inconsistencies, and externally the scientists were being more and more dragged into the economic and political struggles of the time.

### THE RISE OF NAZISM

Until 1933, despite the upsets of the First World War, scientists as such had enjoyed a secure and to some extent privileged position nationally and internationally. Their work for the establishment of truth and the benefit of mankind was supposed to set them above the common conflicts of States and classes. With the coming into power of Hitler they were struck by the first wave of persecution, itself based on a perversion of science which had been used to justify religious prejudices in earlier times. The Nazis, inspired by their racial theories, first struck at the livelihood of Jewish scientists, then at their scientific beliefs, and refugee scientists of distinction appeared in many other countries, carrying with them their valuable learning and also some of the philosophy and prejudices of the German intellectuals.

Twelve years of Nazi power, culminating in a devastating war and the insane scientific slaughter of tens of millions of helpless people, should have been enough to demonstrate to the men of science, no less than to others, the dangers still inherent in the irresponsible greed of capitalism and the need to take steps to prevent their recurrence. But the very enormity of the disasters and the fears for the future that they engendered, powerfully seconded by security and loyalty tests, have had a paralysing effect on the majority of scientists in capitalist countries. They saw themselves as part of a vast machine with the knowledge of what it could do but without the power to arrest its motion. The attitude of conformity, from which only a minority have escaped, cannot be limited to political or economic matters; inevitably it has coloured the character of scientific thought, making it at all points more cautious, vague, and mystical, and above all pessimistic (pp. 1130 f.).

## SCIENTISTS IN THE SOCIALIST WORLD

The attitude of scientists in socialist countries has been polarized by their experiences in a different direction. On the one hand they have suffered from the remorseless devastations of Europe and Asia, which have wiped out the fruits of years of painful effort and sacrifice. Through their experience they have learned something of the frustrated hatred which they inspired among the leaders of the capitalist world. On the other hand they have been inspired by hope, by the capacity for recovery and renewal that the peoples of the devastated lands have shown, and by the sure prospects, given peace, of far greater achievements than before. One effect of this has been to produce a critical, often violently critical, attitude to all aspects of science, theory as well as practice, that seem associated with the destructive and limiting character of capitalism. At the same time it has generated a positive belief in the capacity of the human mind to understand and control Nature that rejects in advance all intrinsic limitations. These attitudes found effective expression in important constructive scientific work. At the same time they had unfortunate negative results. Partly due to the external stresses and partly to certain abuses of the Stalin regime, a dogmatic spirit spread into science in the Soviet Union and in the countries influenced by it. This led to controversies, of which one, that on genetics, is discussed later (p. 957). This had damaging effects on Soviet science and alienated many scientists abroad. The same tendencies led to an over-estimation of national achievements and a corresponding disparagement of the scientific achievements of capitalist countries. These tendencies, however, are disappearing, thanks, on the one hand, to confidence and assurance on the demonstrable achievements of Soviet science as exemplified in the Sputnik, and, on the other, to the greater and more friendly contacts with foreign scientists. Scientific exchanges are multiplying between capitalist and socialist countries on a give-and-take basis, especially in the decisive field of atomic energy. This does not mean that no differences remain. However, they are no longer in the field of science itself, where the appeal of logic and experiment provide for necessarily provisional agreement, but rather in that of philosophic theory, where ideological influences of social and historic origin play a larger part. Though the different experiences of different cultures does in this way give rise to contrasting ideas as to the nature and purpose of science, the conflict between them may illumine the underlying forces of a world in a state of rapid transformation.

## PHASES OF TRANSFORMATION IN THE TWENTIETH CENTURY

The way in which economic and political factors interacted with the development of science will become clearer and more concrete when discussed in relation to the progress of the different branches of science, and this will be attempted in the following chapters. This treatment inevitably breaks up the time sequence, but science has grown so multifarious and is advancing so rapidly that less is lost than would be if an attempt were made to break the period down and, as in previous chapters, to discuss the progress of the whole of science in each period. The events are, however, so recent and so fresh in the memory of most of the readers of this book that it should be sufficient first to recapitulate them very briefly, and then to call attention to them section by section as they arise. This is all the easier in that, perhaps more than any other time in human history, our age is divided up by sharp breaks which mark off very definite phases, each with its characteristic features. The two great wars, with their immediate aftermaths of revolution, block out the early century. They are major events in science as well as in human history.

Before the First World War came, world-dominant capitalism had reached its last stage – the wealthy, peaceful, but increasingly troubled age of imperialism. Between the wars came the establishment of the Soviet Union as a viable economic unit and the great economic crisis of capitalism, with its aftermath of Nazism. After the Second World War, and the triumph of liberation movements in Europe and Asia, reaction gathered itself together and the 'Cold War' was declared. This has already lasted some fifteen years with alternations of acute crises such as those of Korea, Vietnam, Suez, Hungary, the Congo and Cuba, and phases of hopeful but hitherto inconclusive relaxation of tension marked by a failure to achieve even the first step to disarmament.

All through this period, however, has come the rising swell of successful liberation movements of all the peoples of Asia, Africa and Latin America. So far this has been essentially a political movement. Economic freedom has still to be won, and a complex struggle is being waged between the forces of nationalism, socialism and capitalism all over this newly emerging but still largely under-developed world. The strains of this struggle continue to threaten the outbreak of an unthinkably destructive nuclear war between the major powers. We are still living in the shadow of that war and it dominates the work and thought of scientists.

The rapid changes of our time may serve to obscure the fact that we have reached a new phase of the general transformation of society in which the immense newly released constructive forces of a scientific technology are impinging on a world deeply divided and unevenly developed. Nevertheless in the 1960s a new phase of the general transformation of society is definitely opening.

In tracing the interactions of science and society in detail it should be sufficient to have in mind the general character of the different periods, and to remember also that ever since 1917 two world economies have to be considered, and that since 1945 the peoples of Asia and of other undeveloped countries are coming into the picture.

In the following two chapters the progress of the physical and the biological sciences is traced. The treatment is necessarily different in each successive case. The physical sciences, which are discussed in Chapter 10, underwent in the twentieth century a revolution as important and far more rapid than the great seventeenth-century revolution. It was one which enormously increased their power as a means of understanding not only physics and chemistry but every branch of science. Biology, on the other hand, as will be shown in Chapter 11, has been even more deeply transformed through the explanation of biological phenomena in atomic terms, as evidenced by that of genetics in terms of the molecular structure and interrelations of specific nucleic acids and proteins. This closely linked group of discoveries, culminating around 1960, has the same importance in biology as the quantum theory of the nuclear atom had in physics fifty years before. The means of achieving this new vision of biology came largely from outside the discipline of the subject in the form of new techniques, new ideas, and new explanations from other sciences under the pressure of new problems presented by an expanding agriculture and medicine.

# Chapter 10

# The Physical Sciences in the Twentieth Century

## 10.0 Introduction

This chapter is devoted to one vast sector of modern science that can be called very generally the physical sciences, including also the techniques based on them. It is a category better defined by exclusion than enumeration, as one not involving the study of living creatures or of their products as such. For instance, the study of coal as a fuel or as a source of chemical products, properly belongs to the physical sciences; that of the formation of coal and of the light it throws on the conditions in the carboniferous forests belongs to the biological sciences. The unity of the physical sciences is assured by a common quantitative approach to problems, though qualitative description still dominates much of the field of the cosmological sciences: astronomy and geology. That unity, threatened by the dividing tendencies of nineteenth-century specialization, had since been reinforced by the wide range of the new observations and theory of the atom and of quanta. The old major divisions of physics, chemistry, and cosmological science still remain, but they are now recognized as merely practical working divisions; the underlying picture of matter is the same for all. That is why in treating the physical sciences pride of place must be given to the development of atomic physics, both on account of its absolute importance and because its first discovery and subsequent elaboration were made almost entirely in the present century.

The revolution of physics in the twentieth century inevitably introduced a discontinuity between science and technology more marked than at any previous period, and that despite the greatly decreased lag between theory and practice. The basic engineering products, even the relatively novel automobiles and aeroplanes, and the methods used in constructing them, notably mass production, still remain based on the science of the nineteenth century rather than on that of the twentieth. More and more rapidly as the century progresses the gap is closing, or rather it is moving on through the range of industrial processes as the techniques based on the new physical knowledge – first of electronics and later of nuclear physics – penetrate the older industries and create new ones, such as television and atomic power. The existence of this gap, and the active transformation that is going on in industry, constitute one reason why it seems desirable in this chapter to invert the order followed in earlier chapters, and discuss the scientific developments before the technical ones. Another and far more fundamental reason, of which the first is really a consequence, is that in this twentieth century the relations of science and technology are rapidly being inverted. Science is less and less following technology, and technology is more and more following science.

This chapter will accordingly begin with a discussion (10.1, 10.2, 10.3) of the great revolution in physics and of some of its more immediate technical consequences, in atomic energy, electronics (10.4) and solid state physics (10.5). This leads to a discussion (10.6) of the impact of the theory of the atom, and of the new techniques associated with it, on chemistry and on the cosmological sciences. Next (10.7) comes a discussion of the technology of the twentieth century, centred on the motor and the aeroplane, served by an increasingly electrified, mass-production industry and by a scientific chemical industry (10.8), both concerned with the more intelligent exploitation of natural resources (10.9) and with the present over-riding use of science for war (10.10). At the end of the chapter (10.11) an attempt is made to bring out the interconnections of science and technology, to show their relations to contemporary social movements, and to forecast something of what the future holds in store (10.12). The full argument on the constructive and destructive uses of science is deferred until the biological and social sciences can be taken into account (pp. 1299 ff.).

### THE REVOLUTION IN PHYSICS AND ITS PHASES

Nineteenth-century physics was a majestic achievement of the human mind, an achievement that seemed to the people who were carrying it out a move towards a certain completion of our picture of the operation of natural forces on the secure basis of the mechanics of Galileo and Newton. That picture was destined to be shattered at the very outset of the twentieth century and to be replaced by another as yet unfinished. A study of the nature of this revolution can provide important lessons in the internal development of science and in its relations with society.

Though the revolution in physics broke out abruptly – it can almost be dated to a year, 1895 – it has moved forward ever since with steadily increasing momentum, and spread ever more widely through physical science and beyond. It includes moments of unexpected discovery like that of X-rays and radioactivity in 1895-6, of the structure of crystals in 1912, of the neutron in 1932, of nuclear fission in 1938, and of mesons between 1936 and 1947. It includes as well great theoretical achievements of synthesis like Planck's quantum theory in 1900, Einstein's special relativity theory in 1905 and his general theory in 1916, the Rutherford–Bohr atom in 1913, and the new quantum theory in 1925. It is possible, however, to distinguish a great movement underlying these crucial achievements, and to see that this movement did not proceed uniformly, but falls into at least three distinguishable phases, each linked with specific characters of the economic and social pattern.

The first phase, reaching from 1895 to 1916, might be called the heroic or, in a different aspect, the amateur stage of modern physics. In it new worlds were being explored, new ideas created, mainly with the technical and intellectual means of the old nineteenth-century science. It was still a period primarily of individual achievement: of the Curies and Rutherford, of Planck and Einstein, of the Braggs and Bohr. Physical science, particularly physics itself, still belonged to the university laboratory; it had few close links with industry, apparatus was cheap and simple, it was still in the 'sealing-wax-and-string' stage.

Nevertheless, the beginnings of industrial infiltration had already occurred. For example, the great cryogenic laboratory of Leyden University built in 1884 had close connections with the refrigerating industry. The Kaiser-Wilhelm-Gesellschaft Institutes at Berlin-Dahlem were founded in 1911 as an expression of the interest of German heavy industry in scientific research. It was in 1909 that the General Electric Company chose the already distinguished physicist Irving Langmuir (1881–1957) to direct their new research laboratory. It was indeed from such beginnings that the great expansion in industrial science arose.

The second phase, from 1919 to 1939, marked the first large-scale entry of industrial techniques and organizations into physical science. Fundamental research was still carried on mainly in university laboratories, but the great individual scientists now led teams, were beginning to use expensive equipment, and had close links with the big industrial research laboratories. While physicists were working in far greater numbers and disposing of unprecedented wealth, physics itself was beginning to take on a wider range and to show new qualities. It was also beginning to pay off in industry: in radio, television, and control mechanisms. Already in the thirties the influence of war preparation had begun noticeably to polarize physical science. In the service of war, close links were forged between the leaders of research in physics and chemistry and the industrial and governmental research organizations.

The third phase, which though it has run for only a few years, is yet distinguishably different, stems from the even greater expansion of physical science in the Second World War. It is essentially the first phase of government science, carrying with its enormously increased facilities an equally great danger of misdirection and restriction. The expansion of physics can be seen from the figures, which show a rise from 151 to 1,045 in the numbers of honours physics graduates per year from British universities between 1938 and 1962, and from 908 to 6,863 in the number of members in professional grades of the Institute of Physics.

This increase has also meant an ever greater concentration of physical science than in the previous phase. By linking the progress of science directly to that of industry and armaments, it has become, in the capitalist world, more and more predominantly American. Apparatus has become so expensive, and the teams required to work it so vast, that even industry cannot afford them, and only the most powerful states can make significant contributions to physical science. The prospects of older centres of culture are relatively depressed, as they offer few opportunities and cannot compete in attraction for scientific work with the United States. Further, for the first time in history, the association of science with war has split science itself. Secrecy is imposed and with it tests for political loyalty, and science itself inevitably loses any pretence to a politically neutral character.

Between these three phases are interposed the two periods of war science of 1914–18 and 1939–45, which we must consider just as characteristic of the twentieth century as those of the inter-war years. Their scientific contributions were, however, quite different. The war years, particularly those of the Second World War, were above all periods of accelerated and planned application of science. In both cases the future was deliberately sacrificed to the present. An enormous scientific effort went to produce destruction and misery; nevertheless the very success of that effort showed what could be done with the same effort turned to constructive purposes. And both wars, especially the last, provided physical science with problems to solve and the material means to solve them.

## 10.1 The Electron and the Atom

PHYSICS IN 1896

The great movements of the twentieth century, and the revolution in physical science that has accompanied them, have in the course of fifty-odd years turned physics into something almost unrecognizably different. To understand that revolution it is necessary to go back to consider the attitude and status of the science at the beginning of the century. The atmosphere of physics at the end of the nineteenth century was one which combined a coherent and intellectually satisfying theory with increasingly successful practical application. Faraday's and Maxwell's electromagnetism was finding its use in the new electrical light and power networks. The thermodynamics of Clausius and Gibbs was beginning to affect the design of heat-engines and chemical plant. Certainly new inventions were in the air. Electromagnetic theory was to give rise to wireless; thermodynamics had already led to the internal-combustion engine, which was to make cheap transport and human

207. In the closing years of the nineteenth century, public electricity supplies became established; domestic electric light was generally available in London early in 1882. High voltage alternating current station in Sardinia Street, London, of the Metropolitan Electric Supply Company, from *Electricity in the Service of Man* by R. Wormell, London, 1896.

flight possible. All these, however, were but extensions of established knowledge, they offered no promise of leading to anything radically new.

## THE ELECTRICAL DISCHARGE

The change was to come from the pursuit of neglected branches of physics where there were effects not easy to fit into the classical picture, yet apparently so unimportant that no serious doubt was felt as to their ultimate incorporation. Among the first to break the crust of nineteenth-century physical complacency was the study of the electrical discharge. The phenomena of sparks, arcs, and brush discharges had always seemed a vague and unmanageable, though fascinating, minor branch of physics. In the middle of the nineteenth century they had attracted some attention in connexion with the vogue for arc lighting, but this, by the end of the century, seemed destined to give way to the incandescent filament. However, the electric discharge also manifested itself brilliantly in vacua and, owing to the needs of the new electric-bulb industry, there was a drive to improve vacuum technique. As a result both of the revived interest and the new techniques several significant new observations were made in the late nineteenth century. Many of these did not seem explicable in terms of classical physics: Sir William Crookes (1832–1919), in 1876, following observations of Faraday as far back as 1838, observed a luminous glow stretching from the negative end, the *cathode*, of a highly evacuated discharge tube. It seemed to consist of particles of some sort, torn out of the cathode. He called these *cathode rays* a new *radiant* form of matter. This was prophetic, for it was from the study of many such high-speed or radiating particles that the new physics was to be built.

## RÖNTGEN AND X-RAYS

Johnstone Stoney (1826–1911) glimpsed at this possibility and had called the cathode rays *electrons* in 1894; Jean Perrin (1870–1942) showed that they carried a negative charge (1895); J. J. Thomson (1856–1940) measured their speed (1897). In November 1895, the trend of research was abruptly changed by an accidental and altogether unforeseen discovery. Konrad von Röntgen (1845–1923), then an obscure professor of physics at Würzburg, had bought one of the new cathode-ray discharge tubes with the object of elucidating its inner mechanism. Within a week he had found that something was happening *outside* the tube; something was escaping that had properties never before imagined in Nature; something that made fluorescent screens shine in the dark

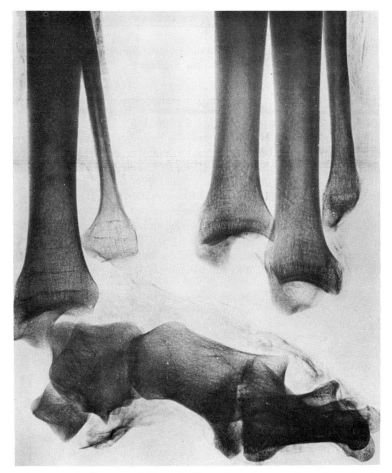

208. Röntgen's discovery of X-rays in November 1895 was almost immediately followed by applications in a wide variety of fields. In 1898 archaeologists were using the technique for examining mummified remains, as this photograph shows.

and that could fog photographic plates through black paper. And they were such astonishing photographs – photographs which showed coins in purses and bones in the hand. He did not know what the something was, so he called it the 'X-ray'. This was a scientific discovery with a vengeance; it was one that anybody could see, and it is not astonishing that within a few days it was stop press news all over the world; it was the subject of innumerable music-hall jokes, and within a few weeks almost every physicist of repute was repeating the experiment for himself and demonstrating it to admiring audiences.

### THE ELECTRON

Great, however, as was the immediate value of X-rays, particularly to medicine, their ultimate importance was much greater to the whole of physics and natural knowledge, for the discovery of X-rays was to provide the key, not only to one, but to many branches of physics. In the first place, it enabled J. J. Thomson to complete his understanding of the generators of X-rays – the cathode rays or electrons – for he found that not only did electrons striking matter generate X-rays, but that X-rays striking any kind of matter generated electrons. They could produce ions or charged particles in gases, and that explained to a large extent the mysterious properties of electrical discharges, including the largest electrical discharge of all – the lightning flash. The discovery that electrons, all apparently identical, could be extracted from the most

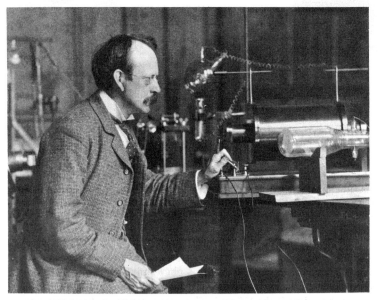

209. Sir John Joseph Thomson at the Cavendish Laboratory at Cambridge, *c.* 1890, surrounded by scientific apparatus of the period.

diverse kinds of matter pointed to electrons as the stuff of electricity. But this stuff was made of individual particles – it was atomic – and it was the consideration of this fact that led J. J. Thomson to take the first decisive step towards the discovery of the inner structure of the atom.

## THE REVIVAL OF ATOMISM

It is in its insistence on atoms as concrete entities that twentieth-century physics differs from that of the nineteenth century. The nineteenth century opened with the atomic theory of Dalton in chemistry. It went on to further triumphs of atomism in the structural formulae of organic chemistry, but, as indicated in Part 5 (pp. 589 f.), the stream of thought in the late nineteenth century, largely under the influence of Mach and Ostwald, was anti-atomistic and was for explaining away the properties attributed to atoms in terms of more general substances and ratios. Newton himself was an atomist, but his mechanics lent itself, when generalized by Lagrange and Hamilton, to a picture of space in which properties varied only slightly from place to place. This *field* type of theory acquired an immense prestige from the intuition of Faraday and its transformation by Maxwell in the electromagnetic theory of light, essentially a theory of fields of force. As we shall see, it was to be generalized still further by Einstein in his theories of relativity.

Continuity was supreme in field physics, which could not easily include the discontinuity of atoms and the even greater discontinuity that was to come with the quantum theory. As in the very beginning of conscious thought about physical phenomena, the idea of atoms had seemed to be a revolutionary one and had always been associated with general atheistic and revolutionary thought. Fields, like perfect geometrical forms, are conservative and continuous. This seemed a much safer kind of physics, but the attempt to re-establish it was a rearguard action that could not hold against the flood of new knowledge interpretable only in atomic terms.

## BECQUEREL AND RADIOACTIVITY

By 1897 atoms had definitely arrived, paradoxically enough by being no longer atoms (uncuttables, p. 183) but by exhibiting a quite disconcerting possibility of being broken up. And not only in the simple way that J. J. Thomson had shown. Simultaneously another discovery of even greater importance had been made. Within four months of the discovery of X-rays, Becquerel (1852–1909) in France, thinking that X-rays must have something to do with the luminosity that appeared in the discharge tubes, tried to find whether other bodies exhibiting a similar luminosity, such as minerals and salts, particularly those of uranium, would show similar properties, and astonishingly enough they did. Here was something like a real accident in the history of science (pp. 607 f.). It was a hint of Henri Poincaré (1854–1912) that had caused Becquerel to ascertain whether there was any connexion between X-rays and phosphorescence. His father had made a magnificent collection of phosphorescent substances. Becquerel might just as easily have picked on zinc sulphide as on uranium nitrate, and the discovery of the phenomena of radioactivity and all it meant for atomic physics might have been delayed for another fifty years. Who knows how many equally simple phenomena capable of revolutionizing our science now lie hidden around us?

The new mysterious rays from uranium were also capable of penetrating matter, and they were produced without any apparatus whatever, spontaneously, from apparently inert and permanent chemicals.

### THE CURIES AND RADIUM: TRANSMUTATION OF ATOMS

This was an even greater shock to the physical and chemical faith of the nineteenth century. The work of the greatest of chemists, of Lavoisier himself, had established the law of the immutability of elements. It had been established as a direct refutation of the claims of the old alchemists to change elements or to create matter; and here

210. The Curies' laboratory in Paris, c. 1904. This illustration from *La Nature*, published in Paris from 1873 onwards, shows a part of the complete laboratory.

apparently was matter actually changing of its own accord without the slightest stimulus to set it off. This was equally a shock to the doctrine of the conservation of energy. Where did the energy which was so apparent in these new radioactive compounds come from? It could only come from within the atom itself. Now an almost infinitesimal amount of radioactive material gave off appreciable amounts of energy. This implied that energy was contained in the atom in quantities quite undreamt of by the users of the energies of burning fuel which was the basis of the industry of the nineteenth century.

Once *radioactivity* was discovered scientific progress was fast – faster, indeed, than in any earlier period in the history of science. Within the short space of six years the essential features of spontaneous atomic change had been laid bare. Pierre Curie (1859–1906) and his Polish wife Marie (1867–1934), the first great woman scientist, itself a portent, had found sources very much stronger than the original uranium. They isolated elements of a new kind such as *polonium* and *radium*, the latter so powerful that it shone by itself in the dark and could inflict serious and ultimately fatal injuries on people who went near it.

RUTHERFORD AND SODDY: RADIOACTIVE TRANSFORMATIONS

Rutherford (1871–1937) had studied the nature of the radiations themselves, and had shown that one type, the alpha rays, were something again quite new in science. They consisted of material particles projected at inconceivable speeds. He showed that the radium atom was giving off atoms, those of helium gas, itself a rare and romantic element first revealed in the sun through the character of the light it emitted, and leaving another atom – that of radium emanation – behind. This was alchemy, natural alchemy; for nothing that anyone could do up to that time could alter the rate of break-up of atoms and their change into other atoms according to set rules of radioactive decay. The pious accepted this as just another inscrutable mystery of Nature and maintained that it would never be possible to interfere with it. With a magnificent combination of physical and chemical techniques Rutherford, now at Montreal and working with the brilliant chemist Soddy (1877–1956), followed up these changes, and in the years between 1899 and 1907 revealed whole families of natural transformations, one from uranium, one from thorium, and one from actinium. Each radioactive element gave out an alpha ray or beta and gamma rays and changed into another, all ending appropriately in the inert element lead. In the study of this process it became apparent that elements were not simple and homogeneous, that each element could contain a number of atoms alike chemically but breaking up physically in different ways. These were the *isotopes* from which so much was to come in later years.

## PLANCK AND THE QUANTUM THEORY

At first this welter of phenomena was so much outside existing theory that they had simply to be put down as brute facts, but already from another part of physics had come a clue which was to help to unravel them. The first discovery of the electron had raised difficulties in the theory of radiation of light. If light is produced by rotating or vibrating electrons it ought to change colour continuously as the electrons lose energy from their radiation; but the plain evidence of the constant wave-length in optical spectra showed that it did not do so. Another contradiction appeared in the theory of heat. According to the classical electromagnetic theory all the energy of a hot body should be concentrated in the short wave-length. It ought to look blue, but it does

211. Max Planck (1858–1947) whose theory, proposed in 1900, that radiation is always emitted in discrete quantities, was seen to fit in well with Bohr's theory of the atom, and was of considerable use to physicists in the decades that followed.

look red. Such discrepancies could not permanently be ignored; but the successful efforts to explain them by Max Planck (1858-1947) in 1900 only got rid of an experimental difficulty to produce a theoretical one. Planck suggested in fact that the energy of atoms could not be given off continuously at all, but came off in pieces; in other words, that energy, like matter, was atomic, but that the atomicity was not in energy itself but in the curious quantity action (or energy multiplied by time). There was accordingly a constant *quantum* or sufficient amount of action, Planck's constant ($h = 6.6 \times 10^{-27}$ erg seconds), that controlled the quantity of all energy exchanges of atomic systems.

### EINSTEIN AND THE PHOTON

Albert Einstein (1879-1955) was the first to draw from this its practical application in the new fields of physics. He explained why it was that the electrons shot off a metal by a beam of coloured light came off at the same speed whether the light was feeble or intense. They could only collect the quantum of energy that the light possessed; more light meant more quanta and not bigger quanta. The speed, however, depended directly on the colour, that is on the frequency of the light. Einstein's picture of the electrons produced by light striking a metal was that of one kind of particle, a *photon* or atom of light of frequency $v$, transferring its energy to another kind of particle, an electron of velocity V or energy E, according to the equation $E = \frac{1}{2} mv^2 = hv$. He had, in fact, reversed the wave picture of light and gone back to the old idea of Newton, that light was made of particles.

### THE ATOMIC NUCLEUS

The full application of the quantum theory to the structure of the atom had, however, to wait for two other crucial discoveries. In 1910 two of Rutherford's co-workers, Geiger (1882-1945) and Marsden, had shown that those natural projectiles, the alpha particles, instead of going straight through thin sheets of matter, were occasionally shot straight back. Rutherford drew from this surprising result (he compared it to a fifteen-inch shell being turned back by a sheet of paper) the simple conclusion that they must have hit something very small and very hard. He had, in fact, seen that atoms had a *nucleus*. This was the other partner of the electron, and because the electrons were negatively charged, the nucleus must have a positive charge exactly equal to the total charge of the electrons around it. How were these electrons arranged? The problem had many curious analogies with that of the arrangement of the planets in the solar system which had perplexed the scientists of the

212. Hans Geiger (1882–1945) and Ernest Rutherford (1871–1937) in the laboratory at Manchester University where they found that alpha-particles were the nuclei
of helium atoms, and that beta-rays were particles travelling at very high speeds.

Renaissance, and it pointed to a similar solution, one indeed which had
been adumbrated by Perrin in 1901, but that could not be proved without facts coming from another quarter: those of the discovery of the
wave nature of X-rays.

### VON LAUE AND THE BRAGGS: X-RAYS AND CRYSTALS

In 1912, von Laue (1879–1960) made the discovery that X-rays could
be *diffracted* by crystals, much as ordinary light was by any fine striated
structure, as by a feather, fine cloth, or a gramophone record, where the
striations have dimensions approximating to those of the wave-lengths
of light (p. 467). X-rays were found to be diffracted by objects of the
same order of size as atoms themselves, and therefore had correspondingly shorter wave-lengths than light. This discovery of von Laue was
as important in its effects as the original discovery of X-rays themselves. In the first place it was taken up by Sir William (1862–1942) and
Sir Lawrence Bragg, father and son, who showed that it was possible
to measure the wave-length of X-rays and at the same time to determine

213. Niels Bohr (1885-1962), a student of Rutherford and a Nobel Prize winner, whose 1913 theory of the atom and its method of radiation of electromagnetic energy was of great significance.

the structure of crystals in terms of the arrangements of the atoms which composed them.

### THE RUTHERFORD–BOHR ATOM

Soon afterwards, in 1913, in Rutherford's laboratory in Manchester, Moseley (1887-1915), a most brilliant young physicist killed at Gallipoli, measured the wave-length of X-rays from a number of different elements, and showed that they followed a very simple law depending exactly on the atomic number or the number of electrons in each kind of atom. Now Rutherford's laboratory,[6.21] owing to the character of the man himself, had already attracted some of the most brilliant minds that had ever worked together in physics. Among them was a young Dane, Niels Bohr (1885-1962), who was able to combine together the four separate strands: the hard nucleus of the scattering experiment, the simple laws discovered long before by Balmer (1825-98) relating to the frequencies in the hydrogen spectrum, the regularity of the wave-lengths of the X-rays from different elements, and Planck's theory of quanta, which would serve to link them together.* Like a new Kepler, he showed that the atom could be pictured as a solar system in which each electron had its own particular orbit, and that light or X-rays were produced only when an electron moved from one orbit of high energy to another of lower energy.

The Rutherford–Bohr atom, the atom of the twentieth century, was now well established in the sense that, as with Newtonian astronomy, it could be used to predict the properties of atoms simply from the knowledge of the number of electrons they contained. It explained the reason why only light of certain frequencies was emitted or absorbed by atoms. Complex spectra could be interpreted and the *energy levels* of the electrons in the different atoms could be found. The very concept of an energy level is a quantum one. It implied that every atomic or molecular structure could exist in a great number of states with different vibration characters like the overtones of a musical instrument and that the *differences* of energy between the states could be found by measuring the *frequencies* of the light emitted or absorbed.

THE NEW ATOM IN CHEMISTRY

But the idea of the Rutherford–Bohr atom could do far more than this. It could be immediately used to interpret the hitherto mysterious and arbitrary laws of chemistry. In the first place it explained why the different atoms had the properties they had; why some formed metals and others did not, and why others again were inert gases. Arrangements with certain numbers – 2, 8, 18, 32 – of electrons seemed particularly stable. If there were more than the set allowed the additional electron or electrons were held much more loosely. In materials composed of such atoms, light set the electrons vibrating easily and was strongly reflected – the characteristic property of a *metal*. If there were fewer electrons than were needed to make up a set, the electrons of different atoms combined so as to share their electrons to the best effect; the result was a non-metallic neutral molecule like those of gases or organic molecules. If *non-metallic* and *metallic* atoms were put together, the metal atom gave up its superfluous electron to the non-metal atom, becoming a positively charged *ion*, and the non-metal *ion*, now charged negatively, combined with it through simple electrical attraction to form a *salt*. In this way the whole picture of the table of the elements, arranged in families and sequences, arrived at logically fifty years before by the great Russian chemist Mendeleev, received a physical and quantitative explanation. There were 92 natural elements, from hydrogen to uranium, because there were elements which had 1, 2, 3, 4, up to 92 positive charges in their nuclei, and each had its own atomic number (p. 568).

### THE STRUCTURE OF CRYSTALS

The discoveries of von Laue and the Braggs were, however, to have other and more extensive consequences. By analysing the relative arrangements of atoms in crystals the Braggs were able to found a new structural crystallography, which in turn was to transform the ideas of the chemists as to the nature of crystals and molecules. It was as if a new microscope had been found which enabled the positions of the chemical atoms to be seen. It could show, on the one hand, that molecules did not exist at all in simple salts like sodium chloride, which were regular assemblies of positive ions of sodium and negative ions of chlorine; on the other hand molecules did exist in such substances as naphthalene, where a group of atoms held closely together were separated by large spaces from other groups – the chemical *molecules* of the nineteenth century. Actually, X-ray analysis was first to confirm and later to refine the structure of molecules, which the chemists had arrived at through a most ingenious mathematical logic based on their transformations into other molecules. Where these chemical methods

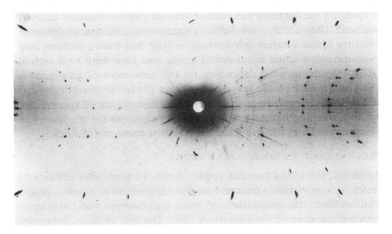

214. By subjecting a crystal to X-ray, a characteristic pattern is formed by the atoms within the crystal diffracting the radiation. The pattern depends on the configuration of the atoms within the crystal and thus provides a means of analysis. The X-ray diffraction pattern shown here is that for sodium niobate (NaNbO₃).

could not be applied, as in the fields of metals and of silicates, X-rays were immediately able to unravel the atomic pattern, and at the same time to account for the special and useful properties of such substances.

## 10.2 Theoretical Physics

### THE FIRST WORLD WAR: RELATIVITY

The progress of physics, after the group of discoveries just referred to, was halted by the First World War, which brought to an abrupt end the first heroic period of modern physics. The war drew some scientists, but by no means the majority, into war service; but even where it did not, it effectively held up, except in neutral countries, the purely scientific research of the non-mobilized experimental scientists. The theoretical scientists for the most part, however, went on working, and it is in that period that one of the greatest advances in the history of human thought took place – the completion by Einstein in 1915 of the *general theory of relativity*. Now relativity belongs in essence much more to the science of the nineteenth than to that of the twentieth century. The keynote of twentieth-century science was discontinuity and atomism; relativity, on the other hand, is still a continuum and field theory; but the fields of relativity are far more generalized than the electro-magnetic fields of Maxwell. They are the new fields of space–time. The special theory of relativity which Einstein put forward in 1905 had shown that, as only relative motion could be observed, space and time were to a certain extent interchangeable, depending on the movement of an observer. Ten years later Einstein was able to bring the hitherto arbitrary and occult force of gravitation into the generalized picture of space–time, but to do this he had to break away not only from the mechanics of Newton, but from the still more firmly based geometry of Euclid.

### EQUIVALENCE OF MASS AND ENERGY

Relativity, for all its popular vogue, is still a theory very difficult to grasp. Its importance in science, however, depends on two closely linked relationships: the equivalence of mass and energy, and the special limiting character of the velocity of light. The first of these, expressed in the formula $E = mc^2$, where E stands for energy, $m$ for mass, and $c$ for the velocity of light, was to provide the theoretical expression of the enormous energies locked in the atom. These were afterwards shown to be the sources of all the concentrated energy in the universe – the energy

215. A total solar eclipse allows stars, which are ordinarily obscured by the glare of the sun's light, to be observed even quite close to the sun's disk. On the theory of relativity, the images of such stars should show a displacement when photographed during an eclipse compared with their positions when the sun was not in their area of the sky; this displacement is due to the gravitational attraction of the sun on the starlight as it passes comparatively close by. This photograph was taken at Sobral in Brazil on 29 May 1919, when the first observational tests of the theory of relativity were made. A star can just be seen between the disk and the mark 's'.

of the sun and the stars, those first nuclear-energy piles. The sun warms us, in fact, by getting lighter, burning up its hydrogen into helium, a form of fire which the successors of Prometheus, undeterred by his fate, are bringing down from heaven in the form of the hydrogen bomb. The limiting character of the velocity of light is an equally significant fact. By showing that all velocities are relative Einstein was also able to explain that, in spite of continuous acceleration, no particle could travel faster than the critical velocity of light, for as it approached that velocity its energy and its mass increased simultaneously so that it became harder and harder to make it go faster.

THE SCIENTIFIC CONTENT OF EINSTEIN'S THEORY

Einstein's theories, for all their abstractness and the fact that they arose from profound cogitation on the meaning of previous scientific theory, were nevertheless derived ultimately from experiments and gave rise to practical applications. The starting point of Einstein's thought was the difficulties inherent in a branch of nineteenth-century physics: the attempt to generalize the electromagnetic theory of light by showing that the apparent velocity of light was dependent on the rate at which the observer travelled through the supposedly fixed ether. This was the celebrated Michelson–Morley experiment, the greatest negative experiment in the history of science. For no difference whatever was found in the velocity of light at whatever speed or in whatever direction the observer was moving. A few years later J. J. Thomson showed that electrons in high electric fields would not move at the velocity they should according to classical Newtonian physics. They seemed to be more sluggish and difficult to accelerate as they moved faster. Both these effects were explained by Einstein's special theories of relativity.

Einstein's *general theory of relativity* went much further. It attempted to include gravitation in the domain of the measurement of space and time. Its particular importance is that it avoided any appeal to what used to be called occult forces like weight or, in more learned terms, gravity, acting at a distance. In their place it postulated that when a body was free, that is not in physical contact with other bodies, it was quite unacted on by forces, and then its mode of motion simply expressed the quality of space–time at the places it passed through. According to this theory our Euclidean geometry applies only to empty spaces – near heavy bodies space is curved. This view marks a return to the original Pythagorean idea of the naturalness of circular motions in the heavens, but it is a return on a higher plane, no longer a semi-mystical intuition but a mathematical account capable of the most refined quantitative proof.

If Einstein had done no more than to find an alternative and neater expression for gravitation than Newton he would have been the Copernicus of the new era; but he did more, he showed that the new method gave results in better agreement with experiment. He was able to explain the apparent shift of the position of stars near the sun by the bending of their rays by curved space and to explain the irregularities in the motion of the planet Mercury. At last Newton's theory of the solar system had definitely been improved on.*

### STELLAR ASTRONOMY AND GIANT TELESCOPES

By then, however, this had long lost the importance that it had first assumed in the days when the orbs of seven planets were assumed to be the steps of heaven. Astronomy had indeed by the twentieth century almost lost both its classical and medieval importance in expressing the divine plan of the world and calculating horoscopes, and its Renaissance importance as a means of navigation. Something of its prestige, however, remained, and this enabled even otherworldly astronomers to wheedle enough money out of hardened business men for the construction of entirely useless telescopes. A giant telescope indeed was the most noble example of the 'conspicuous waste' of Veblen's analysis of capitalism.[6.140a] It showed disinterestedness even more effectively than did shifting European castles over the Atlantic, and retained at the same time the healthy element of competition. Telescopes increased their bore and range by a rivalry as evident as that of the guns of battleships. Whatever their origin, however, the multiplication of observatories with the new tools of photography and the spectroscope brought astronomy far beyond the solar system to the stars and nebulae, which, including our own galaxy, were now recognized as island universes as Kant had first proposed in 1755.[6.124]

### ASTROPHYSICS

The study of the interior of the heavenly bodies as revealed by their light had begun with the discoveries of spectroscopy in the nineteenth century. By the twentieth *astrophysics* was becoming a recognized branch of science, one in which the work of the laboratory and that of the observatory were completely blended. From the outset it had a character different from terrestrial physics in that it revealed structures not only in space but also in time. H. N. Russell's (1877–1957) classification of the spectral types of the stars in 1913 pointed unmistakably to an evolutionary sequence. Cosmology seemed to imply cosmogony; the way things were now could not but raise the question of how they had come into being. In this way astronomy again began to acquire something of its old importance. If it did not reveal the plan of the rational universe laid down once and for all by a beneficent deity, as the Ancients and even Newton believed, it was showing instead an unfolding drama of creation, one which seemed to have some lesson for men. The great development of the knowledge of the history of the universe was, however, to come as a consequence of the further development of nuclear physics (pp. 769 ff.). Einstein had taken only the first step, though it was to be a decisive one. He had shown that the principles of mechanics could be put in question. The quantum theory, in its old and even more in its new form, still further shattered the foundations of Newtonian physics. This revolution was to be as important and as pregnant with further possibilities as had been the overthrow of Aristotle in the Renaissance.[6.130; 6.139]

## EINSTEIN AND THE MYSTIFICATION OF SCIENCE

It is, however, equally true that the effect of Einstein's work, outside the narrow specialist fields where it can be applied, was one of general mystification. It was eagerly seized on by the disillusioned intellectuals after the First World War to help them in refusing to face realities. They only needed to use the word 'relativity' and say 'Everything is relative', or 'It depends on what you mean'. Relativity formed the basis of the work of many popularizations of the mysteries of science.

The physical theories of the twentieth century are no freer than those of earlier centuries from influences derived from idealistic trends from outside science. For all their symbolic and mathematical formulations they still embody much of the flight from reality that derives ultimately from religion, now more and more clearly concerned to provide a smoke screen for the operations of capitalism. The influence of the positivism of Ernst Mach on the theoretical formulation of modern physical theories was a predominating one (pp. 1093 f.).* Most physicists have so absorbed this *positivism* in their education that they think of it as an intrinsic part of science, instead of being an ingenious way of explaining away an objective world in terms of subjective ideas. This was brilliantly exposed almost at the beginning of the period by Lenin in his *Materialism and Empiro-Criticism*; but the mystifications of theoretical physics have still continued, and it will take many more years of argument and experience, including political experience, before the logical basis of physics is cleared of ideas that have nothing to do with the material world.

## EXPERIMENT THE BASIS OF THEORY

The factual history of the development of modern physics shows clearly enough that the advances were, in practically every case, with the significant exception of the prediction of the meson by Yukawa, due to discoveries made in the course of experiments, and that these experiments led to things that had not been conceived of by theory, while the theory was later evoked to explain the experiments. Now the nature

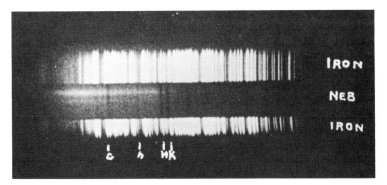

216. The spectrum of a nebula photographed next to the spectrum of an iron spark. The nebula spectrum is in the centre, the iron spark above and below it. The image is a photographic negative so the white spectral lines of the nebula appear black, and the black lines of the iron spark appear white. The photograph was taken towards the latter part of the last century and was one of those that convinced William Huggins (1824-1910) that some nebulae are composed of gaseous material since only material of this kind gives white lines. Huggins's work was followed up especially in the United States, using the large telescopes constructed there at the close of the last century and in the early decades of this.

of theoretical explanation is little more than a language; a physical theory is fully expressed by the equations connecting a set of symbols. The value of the explanation does not, however, lie in the beauty or the simplicity of formulae, but in the number of experimental facts that can be explained by them. That is why the great generalizations of the twentieth century are of such importance. Relativity and the quantum theory cover a far wider field of experience than did the classical theoretical syntheses of the nineteenth century. They have pointed to new experiments which have often proved fruitful. They have, however, failed consistently to explain adequately anything that was not put into them from experiment in the first place.

THE NEW QUANTUM THEORY

The next stage in the history of twentieth-century physics illustrates this most clearly. Bohr's original quantum theory of the atom should, in principle, have explained the structure of all the atoms and molecules. In practice, however, it was found that there was a very awkward difficulty. The quantum numbers attributed to the energy levels in single atoms remained, as the theory demanded, whole numbers, but in the next simplest model, that of a diatomic molecule, the quantum levels of energy starting from the bottom, instead of going 0, 1, 2, 3, very awkwardly went $\frac{1}{2}$, $1\frac{1}{2}$, $2\frac{1}{2}$. This and other anomalies showed by 1934 that there was something very seriously wrong with the form of the quantum theory. It was developing into a kind of formal algebra, almost a cabbala, as it was called in those days, where it was possible to find a set of numbers to explain most things, but not to find any justification, other than convenience, for choosing those numbers. Neither the electron nor the theory of its motion could be as simple as Bohr had originally thought. The first device used to account for this difficulty was to postulate, as Goudsmit and Uhlenbeck did in 1924, that the electron was a little magnet as well as a charge – that it had a 'spin'. Major difficulties, however, still remained.

PHYSICAL EQUIVALENCE OF WAVES AND PARTICLES:
WAVE MECHANICS

The effort to overcome them led in 1925 to a general revision of the quantum theory of a very profound character. That this was overdue is shown by the fact that it was carried out almost simultaneously by four very different physicists: de Broglie in France, Schrödinger and Heisenberg in Germany, and Dirac in England. Their solutions were formally quite unlike each other, though mathematically equivalent. Louis de Broglie in 1923 had followed the track of the history of physics back to the controversy of the seventeenth century between Newton and Huygens[6.32-3] (pp. 466 f.). That controversy had already brought out the striking analogy that, whatever the medium, both particles and waves followed out minimal paths. A wave moved so as to make the *time* a minimum (Fermat's principle), a particle to make the *action* a minimum (Maupertuis' principle). Might not these two principles be reduced to one, thought de Broglie, if particles and waves were essentially identical? Electrons might after all be waves, just as lightwaves might be particles. There appeared indeed to be a general correspondence between particles and waves; every particle could be deemed to be accompanied by a wave and every wave to consist of particles lined up on wave fronts.

Schrödinger in 1926 used this idea to explain Bohr's stationary electronic states in the atom as analogous to the different characteristic modes of vibration of the electrons in the atom, moving not in progressive but in standing waves. This is formally similar to the different characteristic vibrations of a musical instrument with harmonic relations between them. The de Broglie–Schrödinger *wave mechanics* had the advantage of being able to explain the anomalies in the old quantum theory in a way that could be physically grasped as well as mathematically stated. But this was not strictly necessary; Heisenberg and Dirac in different ways scorned even this degree of physical representation. Heisenberg by the use of matrices, or chessboards covered with numbers, and Dirac by an algebra in which $a \times b$ differs from $b \times a$ by $4\pi h \sqrt{-1}$ provided equally good formal solutions to the problems of quantum physics.[6.61]

There have been profound arguments, ever since they were propounded, as to the physical meaning of these theories. Their elegance and their success in explaining facts were for a long time considered to be a complete justification of their truth. However, as time went on it appeared that the new quantum theories, as they were called, were likely to get into as great, though quite different, difficulties, as the old quantum theory. They were able to account for the phenomena that gave rise to them, but as the study of the nucleus and of high-velocity particles progressed new phenomena appeared which were increasingly difficult to account for. A variety of devices and *ad hoc* variations of the quantum theory were resorted to without much success. Nor were the new quantum theories of a sufficiently self-consistent character to be even mathematically acceptable. They still represented an uncomfortable hybrid between the particle physics of Newton, suitably adjusted or broken up by quantum postulates, and an entirely new kind of mathematics, largely determined by statistical considerations. The philosophic difficulties they raised were even more serious.

## THE PRINCIPLE OF INDETERMINACY

Just as in the case of relativity, the new quantum mechanics was in its turn found to be a very convenient basis for mystification. Heisenberg's *indeterminacy principle* was particularly valuable to the reactionary and theologically minded. This states that it is impossible simultaneously to determine with more than a certain degree of accuracy the velocity and position of any particle. Now this as a physical statement is a translation of an equation which is very useful in determining certain observable quantities. The principle of indeterminacy is founded on the success and failure of certain hypothetical experiments. The most famous of these is the gamma-ray microscope, in which the very act of observing a particle drives it from the position it would have occupied unobserved. Useful as illustrations, such experiments, which could never actually be performed, have allowed concepts such as the essential role of the observer, which form no real part of the quantum theory, to be imported into it. As Einstein and de Broglie have pointed

217. Albert Einstein (1879–1955). A photograph taken in later life.

out,[6.32] the attempt to make phenomena subjective in this way leads to paradoxes as formidable as those the indeterminacy principle was constructed to avoid.[6.81]

This principle has, moreover, been given an altogether different meaning by popular scientific writers and even more so by philosophers. Because of this assumed indeterminacy it was claimed that the electron was in a certain sense a free agent. It might or might not at any time do this or that. And if the electron is a free agent why should man not be? Why should not the whole edifice of scientific determinism crash to the ground, to be replaced by a chaos of indeterminacy? Oddly enough, many of the adherents of the new indeterminism were not in fact indeterminists at all. What they wanted was to find a possibility for the interference of God in the affairs of the universe in detail, by slipping the electrons in and out of the places they could occupy in a quite arbitrary manner. The best comment on this was Einstein's, who said, 'I could not respect a God Who spent His whole time in games of chance.'

Actually the construction put on the quantum theory is altogether arbitrary and uncalled for, depending as it does on a particular analysis of the meaning of physical quantity. Even if it were true on the atomic level it would not justify all its extension to the fields of the far more complicated biological and social systems. As we shall see later in this book, the character of physical theory itself had already by the middle of the century come to be as complicated and unsatisfactory as previous physical theories had been before they were transformed by the new outlook. It is important to have in mind the cardinal difference between the theories used to explain and co-ordinate sets of experiments after they have been made, and the ideas which were consciously or unconsciously in the minds of the experimenters who made the new discoveries and opened the new fields to scientific thought.

## 10.3 Nuclear Physics

### RUTHERFORD AND THE MATERIAL APPROACH TO PHYSICS

The great figure of twentieth-century physics, and indeed probably of twentieth-century science, was Rutherford. His work throughout was marked by a simple ruggedness of ideas and an intensely material and mechanical approach to the explanation of physical phenomena. In this respect he resembled Faraday much more than Newton. Rutherford thought first of the atoms, then of the sub-atomic particles he had discovered, exactly as ordinary material particles: as projectiles, tennis, or billiard balls. He treated them as such and found out things about them from how they moved or bounced. Sometimes the particles did not behave as he expected. He accepted the new discovery as a fact and assimilated it by making a new imaginative picture of the structure with which he was dealing. Thus, step by step, he proceeded from the study of unstable atoms of radioactivity to the discovery of the atomic nucleus and the general theory of the atom.

ARTIFICIAL TRANSMUTATION

In his later years he went on to the study of the interiors of atomic nuclei themselves, now merging his work with that of a group of brilliant assistants. In 1919 he made the crucial discovery that it was possible to break up a nucleus of nitrogen by a direct hit from an alpha particle. From now on it was clear that man could control the processes going on in the nucleus if he could find the suitable projectiles with which to attack it. There were two ways of doing this. One was to find among the nuclei themselves those that would naturally emit suitable projectiles, and the other, and more direct way, was to take ordinary atoms and to speed them up by electrical devices.

THE GENERATION OF HIGH VELOCITY PARTICLES

It was this latter method that was first adopted, though paradoxically enough most of the important results were to come from the older methods of radioactively generated particles. Rutherford himself worked with apparatus of a simplicity and cheapness that could hardly have been matched in the nineteenth century and indeed more resembled that of Gilbert in the sixteenth century. This was the famous 'sealing-wax-and-string' school of the Cavendish Laboratory.[6, 94] The simplicity was somewhat fictitious, because in fact the results could not have been obtained without making use of the knowledge laboriously gathered with much more elaborate apparatus in the nineteenth century. Nevertheless, it was in startling contrast to the new requirements for particle accelerating or, as they were popularly called, atom-smashing machines. To get particles up to the high speeds necessary for this required apparatus of a kind different from that which had been found in physical laboratories hitherto, and the construction of these machines meant a new chapter in the history of the relations of physics with industrial developments. Cockcroft (1897–1967) and Walton, with the assistance of the electrical industry, built a high-tension tube through which

218. The main components of C. T. R. Wilson's 'cloud chamber' that generates super-saturated clouds of water vapour in a small cylindrical vessel. With this apparatus the tracks of water droplets that form on water vapour molecules that have been electrified by the passage of atomic particles may be observed. From these tracks the reactions between atomic particles may be photographed.

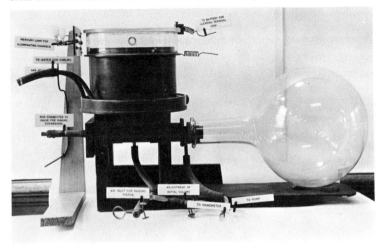

hydrogen atoms could be accelerated with about one or two million volts, and with it showed that such particles could break up the nuclei of a number of light atoms.

### PHYSICS LINKED WITH ELECTRICAL ENGINEERING

The construction of such tubes was possible because of the developments that had been going on in the electrical industry in the earlier years of the century. The need for a study of high-tension lines had come with the increased range of transmission of electric power. At the same time developments in communications engineering, especially the fantastically rapid growth of radio, had led to a mastery of large-scale vacuum technique. The need to construct physical apparatus on an engineering scale meant that from the mid-twenties physical research and particularly atomic research would become even more closely linked to the electrical engineering industry. The expense and the requisite technical experience alone would make it impossible for it to be run any longer as a mere annexe to university teaching. From the two-million-volt accelerator of Cockcroft and Walton have come the host of gigantic

219. An interaction between nitrogen and an alpha particle, causing disintegration of the nitrogen nucleus. Photographed using the Wilson cloud chamber.

modern particle accelerators. The new principle, introduced by Lawrence (1901–58) in the cyclotron, of building up the velocity of the particle not in one burst but in successive impulses opened the way to ever more powerful betatrons, synchrotrons, linear accelerators, to synchrocyclotrons giving the equivalent of tens of billions of volts. The only limit is the cost, which by 1963 had reached the order of £40 million. Already it is beyond the reach of smaller nations, who have had to combine for this purpose.[6.112]

Fully to appreciate these developments, which come later in the story, it would be necessary to consider the growth of another branch of physics – the production and control of free electrons – which is discussed on pp. 773 ff., but to avoid breaking the continuity it is better to continue directly.

NEUTRONS, POSITRONS, AND MESONS

The nineteen-thirties were to witness a new burst of physical discovery as great if not greater than the two previous bursts in 1895 and 1912. Radioactivity, or the study of the atomic nucleus, which had shown little advance in the previous ten years, again became the centre of interest, and gave rise to an unbroken series of experimental discoveries that were to culminate in the control of nuclear processes. The first major discovery was to be that of the *neutron*, produced by bombarding beryllium with alpha particles. Actually, when the neutron was first produced it was not recognized as such, and was imagined to be a gamma ray, just because the concept of a particle that was not charged, which seems simple enough to us today, had, despite Rutherford's prediction of its nature, by then become almost a contradiction in terms.

Once recognized and established, by Chadwick's experiments of 1932, as the proton without its positive charge, the neutron was seen to be the central feature of nuclear structure. Very soon afterwards Anderson discovered another fundamental particle, the *positive electron*. This supplied a needed symmetry between positive and negative in the relations of particles and fitted, far better than did the proton, with its nearly two thousand times greater weight, Dirac's theory that the positive charges in the universe are as it were the missing pieces of a universal negative charge. The relation of neutron and proton turned out to be by no means a simple one. The nucleus which had previously been thought to consist of protons and electrons, was now seen to be better expressed in terms of protons and neutrons, held together by strong forces which Yukawa in 1935 attributed to a hypothetical intermediate particle, the *meson*. This is an example of a fundamental particle first predicted by theory and then observed by Anderson and Neddermeyer in 1936.

Of these particles, the neutron was to prove the most effective in producing nuclear transformation. Because it lacked charge it was able to penetrate very much farther into matter, and to approach and enter the positively charged nuclei of atoms that repelled positively charged alpha particles and protons. In six brief years, from 1932 to 1938, the effects of neutrons on different nuclei were studied. These were years in which science in general and physics in particular were increasingly to feel the impact of the events leading up to the Second World War. The advent to power of Hitler had driven out of Germany and later out of Austria the majority of creative minds in physics. Their work was to fertilize and hasten the development of physics in Britain, France, and the United States while the tightening grip of reaction, obscurantism, and corruption slowed down that at home.

## ARTIFICIAL RADIOACTIVITY: NUCLEAR REACTORS

The first crucial discovery was that of Joliot – that nearly all atoms bombarded with neutrons became themselves radioactive. The logical consequence of this discovery was immense. It meant that natural radioactivity represented only a residuum of the activity of atoms that had not had time to achieve stable states. Already radium had been

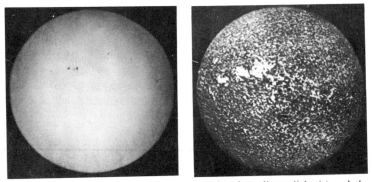

220 a, b. Two photographs of the sun, one taken in ordinary light (a) and the other in the light of the sun's hydrogen gas only (b). The differences between the two pictures permit an analysis to be made of the behaviour of hydrogen and compared with the sun's general features, and so provide a clue to interactions deep within the sun itself.

used to measure the age of the rocks of the earth, and it pointed to the date of origin of the crust as about 2,000 million years ago. But the other elements had been considered more or less permanent. Now this was also put in question, and the knowledge of atomic transformations could be used to explain how the elements had arisen.

## THE HEAT OF THE SUN

This concept was used by Gamov and Bethe to reveal the source of the sun's energy in the mechanism by which four atoms of hydrogen were combined to form one of helium. It was already evident that most of the energy in the universe was derived from nuclear processes. The interest now shifted to precisely how it was liberated. Working up from the light elements, a new nuclear chemistry was appearing, with similar sets of transformations and of stable states to those that had appeared in ordinary chemistry (p. 740). Fermi (1901–54) in 1936 went to the other end of the atomic scale, bombarded heavy elements with neutrons, and claimed that he had produced a number of elements heavier than any that were found in Nature. This he certainly achieved in most cases, but he had, without knowing it, also provoked other changes which were to prove far more important.

## NUCLEAR SPLITTING, 1938

Up to 1937 all the radioactive changes that had taken place had been of the nature of adding small particles to nuclei or removing them from nuclei. The largest fragment ejected was an alpha particle containing two protons and two neutrons. But in that year Hahn (1879–1968) and Strassman discovered that some of the products produced by irradiating uranium with neutrons were of an altogether lower atomic mass, almost half that of the uranium atom. This time it was realized that the atom had been split and not merely chipped, and this knowledge was seen at once to have the most tremendous implications.

Heavy nuclei are able to carry a far greater number of neutrons in proportion to protons than can light nuclei. When the uranium atom split it necessarily liberated several neutrons. Now once this was realized in 1938, largely through the work of Joliot (1900–58), the possibility of large-scale transmutation became an actuality. Here there was a chain reaction or snowball effect. If any nuclear process could be persuaded to yield more than one effective neutron per neutron originally supplied, the reaction would proceed faster and faster. If uncontrolled it would be an explosion, if controlled it would be an energy-producing pile.

CHAIN REACTIONS: BOMB AND PILE

Had this discovery been made in the quieter times of the nineteenth century it would have been pursued ultimately for its practical uses, and possibly after fifty years or so it would have been embodied in new power production machinery. The lack of financial incentives and the vested interest in existing power sources might, however, have held up development indefinitely. As it was, the discovery of nuclear fission occurred on the eve of a new world war. It was fortunate for the British and American governments that among their physicists several, including particularly those who had been driven out by the Nazis and Fascists, were well aware of the military potentialities of the discovery

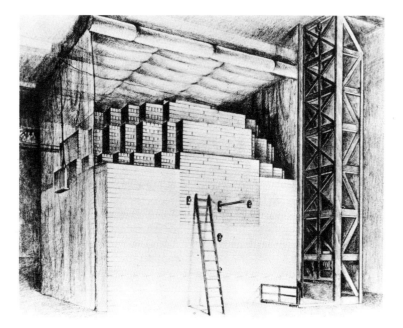

221. The atomic pile constructed at West Sands, Stagg Field, University of Chicago, in 1942, from designs by Enrico Fermi. The device consisted of a pile of graphite blocks in the holes of some of which were uranium. The rate of the chain reaction of atomic fission that the uranium started could be controlled by inserting or withdrawing rods of boron or a similar metal that absorbs the fast neutrons that appear during the fission process. A control rod may be seen close to the ladder. The first self-sustaining chain reaction took place on 2 December 1942.

that had been made. What is perhaps more surprising is that they were able to persuade the military and civil authorities that this was a project worth pursuing with the utmost energy, largely on the grounds that if they did not do so the enemy would certainly get the bomb first. Unfortunately for the German scientists, though luckily for the rest of the world, they did not think the same of the Allied scientists. It seemed to them inconceivable that any scientists other than Germans could ever produce the bomb, and they consequently proceeded in a much more leisurely way.[6.78]

### THE MOST RAPID APPLICATION OF SCIENCE

How the atom bomb was developed, tried out, and used is now part of world history and not merely of that of science. It has been described, apart from its precious 'secrets', in hundreds of books and papers.[6.24; 6.36] Here it is only necessary to say that the guiding physical ideas were derived almost directly from university laboratory experiments and calculations mostly carried out in Europe. The fact that it was successfully developed in the United States was due in part to that country's immunity from actual hostilities, in part to its large available engineering, particularly chemical engineering, resources. This meant effectively that the bomb, and with it all the equipment and 'know how' for the release of atomic energy, was from the outset in the hands of the three or four great trusts of the American electrical and chemical engineering industries.[6.1] This provided an additional reason for the jealous guarding of the secret and for effective reluctance to use atomic energy for power production after the war.

### THE ATOMIC AGE

The military and political consequences of the controlled release of atomic energy will be discussed later. Here it is sufficient to note that technically it represents another major leap forward in man's control of natural forces of the same order as, and possibly of greater ultimate importance than, those of fire, agriculture, and steam. It would appear that this discovery has come only just in time, especially for countries historically dependent on coal, such as Britain, where the rate of power consumption is growing much faster than that of coal production.

Already the cost of nuclear power is comparable with that from thermal sources, and we may reasonably expect that with the use of breeder piles, which produce more nuclear material as they work and can utilize the more abundant thorium as well as uranium as fuel, it will become cheaper as time goes on. There need be no fear for a thousand years or so of any shortage of nuclear fuel. What is holding up the rapid opening up of nuclear energy is, in the first place, the over-riding claims of weapons. Even in Britain, with its desperate need for fuel, all of the new piles that are constructed in the next few years will be producing nuclear material for bombs, and some are mainly devoted to this purpose.[6.87] In second place as a major factor, not so much in construction as development, is the shortage of scientists

222. The nuclear power station at Bradwell, Essex. Heat generated by atomic fission is used to drive steam turbo-generators. Here a charging-discharging machine is shown which can replenish uranium supplies while nuclear reactions are still occurring in the reactor.

and technologists owing to the failure, outside the socialist countries, to appreciate the need for mass higher education in science. Even with these delays, if war can be avoided, the era of nuclear power is rapidly approaching and by the end of the century it will be the main source of electricity.

It may be, however, that within a few decades that power will come not from nuclear fission but from nuclear fusion, or in other words that we will be making slow-burning hydrogen bombs. The achievement of this has proved much more difficult than was imagined at the outset. The problem is one of maintaining hydrogen or deuterium at extremely high temperatures, of the order of hundreds of millions of degrees. At such temperatures matter is in the form of dissociated ions and electrons, so-called *plasma*, and it is obvious that there is no possibility of containing such a plasma in any kind of material vessel, which would be instantly volatilized. However, the very nature of plasma means that the electrically-charged particles in it are affected by magnetic fields and it is possible that some kind of magnetic bottle may be devised which would enable a stable amount of plasma to be maintained in a condition in which thermonuclear fusion can take place. This has largely turned the problem into one of maintaining enormous magnetic fields and here it touches on that of super-conducting magnets involving the maintenance of temperatures almost in the neighbourhood of absolute zero over large volumes.

The problem of thermonuclear energy links together our knowledge of the nucleus with that of astronomy in the generation of energy in the stars (pp. 770 f.). At the moment, however, no one knows – or could tell if he did – how far we are from a thermonuclear furnace or artificial sun. Once this is achieved there will be no need to worry further about energy. We will be able to have as much energy as we can use (p. 851).

While waiting for thermonuclear energy, however, it must be admitted that the prospects of the economic use of fission energy have been disappointing. This is only relative because, effectively, the economics of nuclear power depend on the advances of conventional power sources and generation, both of which have enormously increased in the interval since the first discovery of nuclear fission. What is being sought now is the 'break-even' level between the real cost of nuclear and conventional energy per unit. This is a largely artificial figure which depends on many other factors than physics and the hopes of nuclear energy in countries with little conventional fuel have to be damped by the fact that it is just those countries that are in no position to indulge in large capital expenditure.

Multiple discoveries of sources of oil and natural gas and the increased capacity for moving both of these by tankers or by pipe-lines mean that we are not so dependent as we thought we were on the development of nuclear energy. Nevertheless, with energy requirements continually rising it seems clear that the 'break-even' point will be reached and passed before the end of the century, after which the nucleus in one form or other will become the main source of energy.

By-products of nuclear energy production are already being of use to science and humanity. Among them are many radioactive isotopes (pp. 735 f.), effectively labelled atoms available to match most of the hundred odd elements that exist, or can be made. These tracer atoms readily reveal themselves by their radioactivity, and thus very minute quantities of them may be used to follow the kinds of combinations and dissociations that atoms go through in chemical operations, including the chemical changes which occur inside living organisms. Other uses of piles and pile products are as substitutes for expensive radium and as promoters of polymerization and hardening of plastics.

COSMIC RAYS AND FUNDAMENTAL PARTICLES

Another weapon of still greater power but so far without military application is furnished by the study of *cosmic rays*. These were

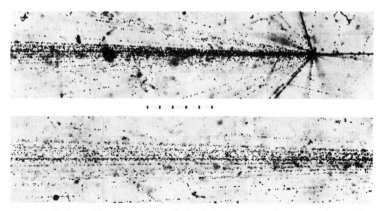

223 a, b. By launching an unmanned plastic balloon carrying a container with a pile of photographic plates (a) the results of cosmic ray bombardment may be recorded on the photographic emulsion. A heavy fast-moving nuclear particle (b) strikes a nucleus and disintegrates, emitting alpha particles as a narrow jet. Plate 223 b is from *The Study of Elementary Particles by the Photographic Method* by C. F. Powell, P. H. Fowler and D. H. Perkins, Pergamon Press, London, 1959.

discovered nearly fifty years ago by the just detectable effect they had in discharging well-insulated bodies. Step by step their origin in the outer universe and their highly penetrating nature were recognized. New techniques, based on an examination of the tracks of individual particles, in cloud chambers by Blackett and Skobeltzyn and in photographic plates by Powell, have revealed a variety of particles some of which are so energetic that they not only penetrate or split atomic nuclei but also cause them to explode into many fragments.

From these studies it appears that the electron, the proton, and the neutron are not the only elementary particles or nucleons but merely the stable or long-lived ones. There are as well a very large number of unstable intermediate elementary particles, the *mesons* (pp. 755 f.). The study of these particles, now becoming known as the fundamental particles of physics, became in the early sixties a headlong race. All attempts to limit their number either by a comprehensive theory or by exhaustive experiments have been broken down, usually within a few months, by the prediction or discovery of a new particle or group of particles. Already they are to be counted in scores. I would hesitate to say how many there would be by the time this volume is published.

Nevertheless, the fundamental particles are now beginning to show a definite pattern. Each is accompanied by what is called an anti-particle, as the positron is to the electron. When two such particles meet both disappear by mutual *annihilation* and their energy is transformed into a pair of photons. They can also be *created* as pairs in other energetic collisions of other fundamental particles. This shows how relative are our conceptions of material existence which hold only in our familiar world of low energy exchanges.

However, the pattern they make seems to indicate that they might equally well be taken as different *states* of the same basic energy concentrations or mathematical singularities of some field deeper than that of the electro-magnetic or the conventional meson fields. Many, perhaps the most interesting, are extremely short lived, that is of the order of $10^{-27}$ of a second, hardly long enough for a particle moving with the speed of light to go through more than a nucleus. The study of fundamental particles is now the most exciting field of pure physics, bordering on philosophy and mathematics. The smaller and more ephemeral the particle, however, the larger and more expensive is the equipment required to study it. Here at centres such as Brookhaven in the US, Dubna in the USSR and CERN in Geneva, theory and practice go very closely together. Apparatus is designed and built, costing millions of dollars, containing thousands of tons of concrete and steel, to test the

224. Nuclear disintegrations result from the interactions of nuclear particles in the modern accelerator like the synchrotron. Often a protracted search of experimental records is necessary to discover a particular interaction. The discovery of the anti-sigma-minus-hyperon here recorded was only found after screening and analysing 40,000 separate photographs at the international nuclear research laboratories at Dubna in the USSR.

efficacy of a piece of pure mathematics, often with startling and un-expected results. They have enabled the physicists to penetrate into the structure not only of the nuclei but even of the nucleons, the protons and neutrons of which they are composed. It would appear that the nucleons have themselves an inner nucleolus surrounded by a cloud of mesons.

### NEUTRINOS

The smallest and strangest of the fundamental particles are the neu-trinos, without mass or charge, which we now know occur in four forms, those associated with the decay of electrons and mu-mesons and their anti-particles. They interact very feebly with other particles so that most can pass right through the earth without deflection and can be detected by their going upwards into the apparatus. First predicted theoretically by W. Pauli (1900–58) in 1928, they have now been de-tected, but only by very tedious and expensive experiments which cap-ture only one out of many billion particles. It may well be, however, that despite their minuteness and inertness they may have a large part to play in the evolution of galaxies and in the creation of matter.

### NON-CONSERVATION OF PARITY

Another breach in our common-sense view of the world was the dis-covery by Lee and Yang in 1960 that for certain strong interactions of particles right- and left-handed spins did not occur in equal numbers, which means that in the universe – at least in our part of it – there was a definite intrinsic twist which may or may not be connected with the twist or molecular asymmetry of living organisms discovered by Pasteur (pp. 629 f.). Symmetry of rather an odd character remains. Parti-cles and anti-particles spin in opposite senses. The complete reflection of a world would be an anti-world.

### THE PRINCIPLE OF UNITARY SYMMETRY

In 1964, the first intelligible account of the arrangements of the so-called fundamental particles was reached and a principle, called the Principle of Unitary Symmetry, was evolved as a result of a remarkable international effort, which began with the Japanese, Ohnuki, in 1960, and was taken up later by Salam from Pakistan, Regge from Italy, then Ne'eman from Israel, and later by Okubo, again from Japan, and Gell-Mann from California.

They were able to show that the baryons, or heavy unstable particles, twice the mass of a proton, have so-called strong interactions. They are

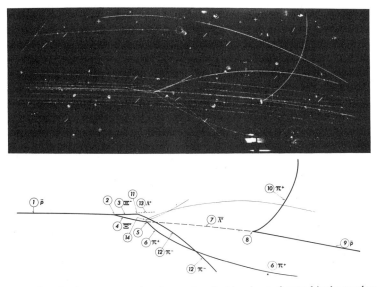

225. Another important nuclear interaction that has been observed is the production of the anti-cascade particle known as the positive xi-minus. It is caused by the action of an anti-proton (a negatively charged proton) with the normal positively charged proton. When these two particles collide, they suffer mutual annihilation and out of this come an anti-particle, the xi-minus (written Ξ⁻) and an anti-anti-particle, the positive xi-minus (written $\overline{\Xi^-}$). The existence of the positive xi-minus was predicted theoretically but observed only some years later when experimental techniques were sufficiently developed. The xi-minus and positive xi-minus are marked in the diagram below the photograph; the other Greek letters refer to other nuclear particles. Photograph by C E R N (see plate 262).

set in a pattern, a set of patterns of eight or ten members, according to various combinations of their isotopic spin or hypercharge. This apparently purely formal arrangement gave a prediction of the properties of an undiscovered particle, the Ω minus baryon of 1,685 electron masses.

A particle of precisely these properties, and a mean life of $10^{-10}$ seconds was subsequently discovered at Brookhaven. The theoretical physicists have at last shown that they have a method of prediction of particles and their properties, though they are not as yet able to explain the physical theory behind their predictions, in the way, for instance, that the Bohr theory was able to explain the Mendeleev Periodic Table in terms of the outer electronic parts of the atom (pp. 739 f.). The discovery of the predicted $\Omega$ minus particle is felt to be of enormous importance because for the first time it breaks the block which has held up theoretical particle physics ever since the appearance of the new quantum theory in the twenties.

We can see now that we are not so much dealing with entities as with systems of forces, or interactions, of which we already know four and of which the strongest, or nuclear force, has a range of the order of $10^{-13}$ cm., the diameter of a nucleus with energies running into tens of electron volts.

The next strongest has long been familiar as electromagnetic force, which is about one-hundredth as strong. Unlimited in range, this is the force that binds the electrons to the nuclei and makes neutral atoms. Next is the so-called weak nuclear interaction, only $10^{-14}$ of the strength of the strong interaction, which again has very short range and is responsible for the decay of nuclear species and the production of leptons which include the photon, electron, the mu-mesons and the neutrino. The weakest force of all is the force of gravity, only $10^{-39}$ of the strong interaction. But this force is effectively also widest in its range and dominates the shape and movement of large masses in the universe. All that physics has done so far is to classify the nature of these forces and their interrelations and their full significance is a matter for the future.

The existence of short-lived particles shows that our ordinary experience of the world is only a very limited one, limited by our own capacities for apprehension. Many things exist and may play an enormously important role in Nature, but are not revealed to us either because they are too small or because they are changing so rapidly. Everything we consider permanent merely corresponds to a long-maintained stage in a sequence of changes, and the elements of the Victorian scientists, like those of Heraclitus, are in a state of perpetual flux. The flux may not always be moving at the same rate. There is considerable evidence that the great majority of the elements we know today on the earth were built up by processes of the same kind, but much more powerful than those that go on in atomic piles. The mere fact that they exist, and their relative abundance or rarity, furnishes evidence for deducing the circumstances of the original formation of the solar system and the planets some 6,000 million years ago.

THE NEW COSMOLOGY

The advance in our knowledge of and interest in the universe has suffered an explosive change in the last two decades. It is the example of a great combined operation by which the original optical study of the universe and the evolution of astrophysics, already mentioned (p. 745), has been added to by the advances in the knowledge of nuclear structure and the vast amount of new information provided by radio-astronomy (pp. 779 f.).

Since 1957, the beginning of the space age, these indirect methods of exploring the universe have begun to be supplemented by the more direct approach through actual space travel. In the whole field of astronomy the findings of these different branches have now come to

226. Galaxies form the main 'units' of the cosmologist, who seeks to determine both their past positions and original formation. Each galaxy is either a conglomeration of stars, dust, and gas and has a spiral shape – like this galaxy in the constellation Canes Venatici – or seems to be more amorphous and composed primarily of stars alone.

reinforce each other and have led to a great break-through in our understanding of the universe, a new cosmology which is still in the making and advancing as rapidly as our knowledge of nuclear physics. It is, correspondingly, the more difficult to set down on paper results which will be obsolete even as they are being written. Here cosmology, the description of the universe, has now been necessarily compounded with cosmogony, the history and development of the universe; one is not understandable without the other.

In the first place we may well ask 'What is the universe?' This has become almost as relative a matter as it was in the days when our savage ancestors saw gods and heroes in the sky, with the chariot of the sun sweeping through it by day and of the moon by night. The universe is as much as we can at the current moment see or find anything out about. By now the achievements of the giant telescopes of the pre-war decades have been enormously outranged by that of the even more gigantic and strange radio-telescopes; but nothing yet points to any kind of limit or ending – the further you can see, the more there is to be seen. Yet all those paradoxes still remain. If the universe is strictly infinite and full of luminous objects, it should provide a sky not covered with points of light but uniformly bright. Things cannot be as simple as this.

Part of the key to this mystery is provided by the observation, already an old one, that the universe appears to be expanding, that is, the galaxies move away from us more rapidly the further out they are. For a long time this view seemed to imply that at a time, not so terribly distant, which would vary from eight to twenty billion years, all the galaxies of the present day were very much closer together. The suggestion, which was that of Lemaître as far back as 1927, was that the universe started with a kind of cosmic egg or, as we would now think of it, cosmic atom bomb, which burst, scattering its expanding parts all over empty space. They are still moving and separating still farther apart. This is a return, on scientific grounds, to the old cosmogony of the universal egg. Naturally, it has not gone unchallenged.

The most effective challenge to this single creationist theory produced another, equally difficult to accept, to our naive view of the way things happen. It is that matter is continually being created in the various spaces of the universe and coming together into galaxies which then separate and end up in majestic explosions scattering the seeds of matter all over the universe. This is the view of F. Hoyle and H. Bondi, which eliminates the necessity for a special start to the universe but puts in its place something more difficult to visualize, that of a process which has no beginning and no end.

227. Galaxies appear for the most part in clusters which may contain anything from a dozen or so to hundreds of separate units. A cluster of four galaxies in the constellation of Leo. The stars (dots) are comparatively close to us and are part of our own galaxy. The galaxies themselves lie at such a distance that their light has taken some one and a half thousand million years to reach us.

That the universe is full of cataclysms is now becoming more and more apparent. Some, at least, of the most important radio sources that were once thought to be quite close in our galaxy are now known to be a prodigious distance away and come from an explosion capable of destroying not merely stars but clusters of galaxies. More recently still some of these objects were found to be quite small, more like stars than galaxies, enormously energetic, equivalent to some billion suns. They may be either very far away or enormously heavy to account for their large red shift. This shows that the universe may contain many other kinds of objects unknown to us and not predictable by existing theories (p. 780).

Stars themselves are known to evolve and to be born at very different epochs because stars containing short-lived radioactive atoms, as revealed by their spectra, can be observed in the dustier portions of the galaxies. The question of the *evolution* of the universe is absolutely tied up with its structure. Indeed, the galaxies that we see at the greatest available distance are, or were when they first emitted the light or radiation that reaches us, something between five and twenty thousand million years old. We are literally looking into the past. Nevertheless, at the moment observations, experiments, and theories are in such a state of flux that all that seems established is that the universe has a history.

THE INSUFFICIENCY OF PHYSICAL THEORY

In tracing it out as much is likely to be learned about the nature of matter and radiation as about the distant heavens. Indeed the new discoveries, especially those of the fundamental particles and their transformations, have put a very considerable strain on existing physical theories, particularly those of the laws of interaction of elementary particles and of the constitution of the nuclei. Such theories as exist – and it must be admitted that for many phenomena there are no theories – are built on *ad hoc* analogies with the quantum theory applied to the much stronger forces and smaller distances in nuclear physics. Involving 'cloudy crystal ball' models, 'magic numbers', and 'strangeness' quantum numbers, they even have a somewhat magical – cabbalistic – flavour.

It may well be, however, that a far more radical revision of the relativity and quantum theories needs to be made, not by tinkering with the present theories while accepting the assumptions that underlie them, but rather by making a fundamental attack on their logical and philosophical bases. It was in this way that the older theories were overthrown, first by the accumulation of material experimental evidence which they could not explain, and secondly by questioning the bases of the arguments which had led to the classical theory. Any new theory must of course account for all or most of the existing facts, but it will be accepted only if besides explaining them it serves to link together more successfully even wider fields of experience.

We are just entering a new phase of criticism of physical theory where the evident *malaise* of mathematical physicists at the inadequacy and inelegance of the quantum and relativistic theories is giving rise to efforts at radical reconstitution. Though the new theories are various they have common aims. One is to generalize a field theory that will unite the hitherto disparate relativity and quantum theories. Another is to remove the need for the basic indeterminacy of the new quantum theory of 1925 especially associated with Bohr and Heisenberg. The victory will go to whoever can explain satisfactorily the new and fuller range of physical phenomena, the intranuclear forces, and the behaviour

of the range of ephemeral and protean particles. It is too early to say
what will ultimately emerge, but it is bound to be very different from
the accepted orthodoxy of the last forty years (pp. 861 f.).

## 10.4 Electronics

WIRELESS AND THE IONOSPHERE

We have here pursued the subject of nuclear physics to the bounds of
present knowledge. But nuclear physics, though it represents the farthest
outpost of the advance of experiment and theory into the unknown, is
not the whole of physics and not even the most useful part of it. Indeed
it could not have come into existence if at the same time great advances
were not being made in other fields of physics. The most important were
in the fields of radio waves and electronics. Here the development of
physics ran parallel with that of industry. Electromagnetic waves had,
as we have seen, been produced by Hertz in 1886 following Maxwell's

228. A photograph taken about 1901 of Guglielmo Marconi (1874–1937) (left) with
his assistant George Kemp. The photograph was made during a successful trans-
atlantic radio experiment.

theory on their nature and properties. It was not until the end of the century that they were used for practical signalling. By then the interest they aroused induced successful trials in many countries; by Oliver Lodge (1851–1940) in England, Popov (1859–1906) in Russia, and Bose (1858–1937) in India, among many others. Full commercial success did not, however, here go to the trained scientist but to the gifted and optimistic amateur.

A sound physicist would have said at the beginning of the century that it was quite impossible to send electro-magnetic waves over any large distances. They would simply go off the surface of the globe through the air and not come back. Nevertheless Marconi (1874–1937), who was not enough of a physicist to believe this, tried to send wireless signals across the Atlantic and they were actually received on the other side. This meant that there must be some kind of mirror which reflected radio waves back again down on to the earth. Appleton (1892–1965) took up this study in the twenties and was able to show that such layers, consisting of ions produced by solar radiation, existed not only at one but at several levels in the atmosphere's constitution – in what is called the *ionosphere*. He measured their height by sending up very short signals and noting the time they took to be reflected. This was the basis of the *radar* device of the war, essentially the same as the method of echo-sounding which had already been used in the First World War for locating submarines by the very much slower movement of pressure waves in water, and indeed the method used by bats in avoiding obstacles in the dark.

## THE ELECTRONIC VALVE

Marconi's spectacular and unexpected success assured the rapid development of wireless communication if for no other use than for communication with ships at sea. It would not, however, have taken the place it has in everyday life had it not been for the development of the electronic valve. This major contribution to twentieth-century electronic physics came almost equally from industry and science. Its transformation from a laboratory curiosity to a saleable commodity in less than a decade is a measure of how rapidly industry could absorb and utilize twentieth-century physics. The initial observation which led to the development of the valve came from industry itself, indeed from Edison's own research laboratory in Menlo Park. Already in 1884 he had noticed that the glowing filament of an electric bulb could retain a positive but not a negative charge. He sealed a metallic plate into the bulb and found he could pass a current from plate to filament but not from filament to plate. This was the first *electric* valve, and its action was readily explained by J. J. Thomson's theory of electrons. The hot wire of the filament gave off electrons which travelled to the plate only if it was charged positively, but the cold plate could not give them up even if charged negatively. The dependence of the valve on the properties of electrons justifies its modern name – the *electronic* valve. The two-electrode valve was found useful as a rectifying device in radio telegraphy. It was, however, modified somewhat empirically, by de Forest (1873–1961) in 1905, by adding another electrode in the form of a grid, to make the three-electrode (triode) valve, which gave it the really

229. The first electronic valves were constructed at University College, London, by Ambrose Fleming (1849–1945). They were diodes (valves containing two electrodes) and were used for rectifying radio waves to assist in their detection.

revolutionary possibilities of amplification and generation of waves. This device made radio telephony and broadcasting possible, and is the basis of all high-frequency engineering today, both in radio and to a larger and larger extent in power electricity.

AMPLIFICATION AND REGENERATION

The *triode valve* and its numerous and complicated progeny are not merely or even essentially valves. Its real novelty is that it is an amplifying device; it allows small variations of voltage or current to be converted into large ones. The principle of *amplification* is that small energy changes can be made to direct large ones. Earlier devices, such as the lever, had magnified mechanical action, or, like the lens, had spread out images, but in all these cases the applied energy was merely transmitted and some was always lost. In the amplification effected by a valve, energy is fed in from outside but the pattern can be imposed on it by one that is much weaker. The valve is the type of device operating on *information* rather than on power. It was indeed the first fully flexible *cybernetic* device (pp. 783 f.) – an enormous step from its crude anticipation in the escapement of a medieval clock or the electrical relay of the nineteenth century. By coupling the output of a valve back on itself in a resonant circuit it can also be made to generate oscillations of controllable frequencies. These two properties, *amplification* and *regeneration* or *feed back*, made the valve at the same time an observing instrument and a tool. It is perhaps the most characteristic product of twentieth-century technology. It is the same functions, rather than the particular way of achieving them, that characterize the smaller and more versatile successor of the electronic valve – the transistor.

The development of valve manufacture found its basis in that of electric lamps and in turn the more severe demands made on valves stimulated vacuum technique. It was enormously stimulated by the use made of valves for radio communication in the latter years of the First World War, and soon after it by the new popular demand for radio. Now, once it could be manufactured cheaply and on a large scale, the valve could return to the service of physical science. Indeed, it is impossible to imagine how physical science could have achieved the results it did in the second quarter of the twentieth century had it not been for the universal use of valves, which could have been made sufficiently cheaply only by their having an important industrial use. The developments of high-tension, vacuum, and valve techniques inevitably led to an integration of academic physics and the electrical industry in the twentieth century as close as that which existed between academic chemistry and the chemical industry in the nineteenth. A new applied science was born and acquired the very appropriate name of *electronics*.

RADIO AND RADAR

Its first uses were in the refining and extending of radio communication. There was a steady trend to shorter and shorter wave-lengths, partly because of the exhaustion of available bands through the ever-increasing number of broadcasting transmitters. Another advantage of shorter wave-length was the increased possibility of directing it along well-defined beams. Directional radio started from the need to detect the origins of the thunderstorms that caused the troublesome atmospherics, and it was later used for beamed wireless for distant sending. Accuracy in direction, however, depended essentially on using shorter and shorter

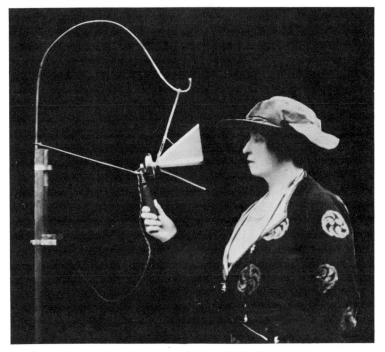

230. In 1921 regular broadcasting was initiated in the United States and concerts were broadcast from The Hague. Late in 1922 the British Broadcasting Company commenced operation but, before this, the Marconi Company had made transmissions from time to time and on 15 June 1920, Nellie Melba made a broadcast from the Marconi Company's experimental high power station at Chelmsford.

waves, and this in turn reacted on the manufacture of the valves and circuits used to generate them.

From directed waves it was natural to pass to the study of reflection and hence to radar. The immediately effective stimulus for its practical development lay in the threat of air attack that hung over the world before the Second World War. Once the problem of detecting the presence of an aeroplane by the reflection of a pulse of radiation was formulated, it was not long before intensive and organized research led to an effective solution. In Britain, thanks to the initiative of Watson-Watt, a radar screen was developed just in time to check air invasion in the second year of the war.[6.127] Soon after a further great advance was made in the invention of the cavity magnetron as a powerful source of centimetric waves, enabling much higher precision of location to be achieved. As the war advanced radar came to be used in an ever larger number of applications: for finding the way, for mapping from the air, for controlling the flights of aeroplanes, and then that of bombs and shells.

231. Experimental work on radio 'noise' and 'atmospherics' by Karl Jansky for the Bell Telephone Laboratories in New York led to his discovery of cosmic radio waves. Although this aroused some interest, little notice was taken by astronomers until after the Second World War when radio astronomy became established as a separate science. Illustration from a negative supplied by Karl Jansky.

232. In 1958 the spectacular radiotelescope at Jodrell Bank was completed. With a reflecting surface 250 feet in diameter, it is still the largest steerable radiotelescope in the world.

### SHORT WAVES: RADIO-ASTRONOMY

At the end of the war short-wave and ultra-short-wave wireless equipment was in common production – a development which again would have taken many, many years under peacetime conditions – and with these short waves man has acquired a new kind of sense organ, one more suited to long and middle-distance observation and communication than anything that ordinary light can give. While by ordinary optical methods only the direction and character of a distant signal can be gauged, radar provides the additional co-ordinate of the distance. Thus it is possible to use these new methods for astronomical purposes, providing a useful check, for instance, of the distance of the moon. More surprisingly it emerges that the sun and stars themselves emit rays of this kind, and these rays give rise therefore to a new kind of astronomy, *radio-astronomy*, showing the existence of invisible stars.

Many of these turn out to be at distances far greater than optical astronomy can reach. The great dish radiotelescopes like the first of Sir Bernard Lovell's at Jodrell Bank, have proved to be as expensive as particle accelerators. They have now found a second use in tracking satellites and planetary probe vehicles.

The link between radio-astronomy and optical astronomy is getting closer. Distant objects such as 3.C. 273, which were first detected by their strong radio signals, have now been verified by optical evidence. They are very strange objects indeed, apparently of stellar dimensions but of the same weight as a whole galaxy and moving very fast, about a sixth of the speed of light away from us, and, therefore, at an almost incredible distance. The quasi-stellar radio sources give us an additional and unsuspected evidence on the origin of the galaxies. It is clear that we are only beginning to understand the processes by which our universe was formed and maintained (pp. 771 f.).

ACCELERATORS AND MAGNETIC RESONANCE

The development of ultra-short-wave radio made possible unexpected applications. The linking of electro-magnetic waves with the movements of electrons can occur in a number of ways and it can be used to project electrons and consequently to develop some of the high speed accelerators and synchrotrons which are the basic tools for modern nuclear physics (pp. 752 f.). Electrons can, as it were, be made to ride on the radio waves pushed along suitable guides. Alternatively, they can be made to swirl around in a magnetic field, either linked to molecules or to the nuclei of atoms. Thus have arisen two new aspects of electronic physics, electron spin resonance and nuclear magnetic resonance, similar in many respects to spectroscopy in their accuracy. Both can be used to characterize molecules and have become one of the most important tools in modern chemistry (p. 796).

CATHODE-RAY TUBES AND TELEVISION

From the early experiments of J. J. Thomson onwards, moving beams of electrons had been used in various modifications of cathode-ray tubes to analyse rapidly-varying currents by transforming them into visible moving images. The cathode-ray oscillograph is in itself a kind of time microscope capable of following changes far more rapid than any system of mechanical levers or mirrors. Its uses in science and industry are manifold. It is now familiar to millions as the television screen. In television, moving electron beams are used in the transmitter to scan electric charges produced photo-electrically from a lens image. The resulting pattern is reproduced by another synchronously scanning beam to impress the fluorescent screen in the receiver. The development of television was slow not because its principles were not grasped at an early date (Campbell Swinton's proposals on essentially the same lines as are now used were made in 1911) and not because of the technical difficulties of scanning or of broad-band short-wave transmission. It lagged essentially because the big electrical firms, even the new firms that had grown up with radio, were too intent on immediate profits to indulge in expensive development. It was left to enthusiastic amateurs like Baird (1888–1946), using primitive equipment, to make the decisive advances and convince the commercial world that there was money in it.

Television, though the most direct of cathode-ray display systems, was not the only one. The needs of the war, particularly that of seeing without being seen, gave rise to many others. The great range of receptors, scanning and transmitting circuits and displays has now made it possible to take any kind of initial radiation – X-rays, ultra-violet, infra-red, or short-wave radio – and use a cathode-ray tube to build up an image visible to the eye. The importance of this in enlarging human perception is especially great because the human brain is itself more than half taken up with the process of seeing and interpreting what is seen.

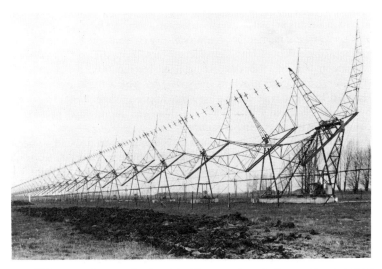

233. Using more than one fixed aerial with open wire reflectors, radio waves from space may be made to interfere and, after analysis, permit of great accuracy in determining the position of a radio source. This radio interferometer aerial at the Mullard Observatory, Cambridge, extends 1,450 feet in an east–west direction. The rotation of the Earth allows it to scan the sky while the aerial can be adjusted to point in a particular north–south direction: with a number of sweeps taking weeks to complete the whole northern hemisphere of sky can be observed.

The eye-brain complex is, as Wiener (1894–1964)[6,146] has pointed out, itself an extraordinarily compact and efficient nerve circuit for recognizing, analysing, and following images. To make a phenomenon visible is immensely to enlarge our powers of understanding it.

## ELECTRONIC PREDICTORS AND SERVO-MECHANICS

Another unforeseen by-product of the development of radio engineering in the war has been the development of electronically linked combinations of receptors and servo-mechanisms, realized in predictors and later in computing machines. These were primarily used for aiming, steering, guiding, and exploding weapons, ranging from a radar-controlled system of anti-aircraft guns to the millions of electronic proximity shells that they fired. This had added a new dimension to mechanical production. Just as the tool is a substitute for the claws or teeth and the machine for the arm and body manipulating the tool, so is the electronic servo-mechanism a substitute for the whole man—eye, brain, and hand together. It is an extension of automatism from a regular routine nature, for which the old machine is adequate, to one in which variations within very wide tolerances can occur.

A servo-mechanism must contain sensory elements such as photo cells and motor elements such as electric motors. It must also contain some connexion between them involving fixed instructions, conditional instructions, and even previous messages, by which the various stimuli received by the device are led to effect appropriate external responses by means of circuits which will be discussed more appropriately in connexion with electronic computers.

By combining valve circuits in various ways, it was possible to begin to make use of the extremely light and flexible character of electronic movements for many of the purposes for which human thought has been needed in the past. What has been done is effectively to increase the speed of all operations of a significant rather than a massive character by a factor of several hundreds of thousands, that is, to do in a ten-thousandth of a second what used to take a minute by mechanical means owing to the intrinsic inertia of massive matter.

At the same time it is also possible to compress into an extremely small space electrical circuits which, if replaced by mechanically operated parts, would take up many thousands of times as much room. Even now this process of *miniaturizing* is only beginning. Newer developments make it certain that this process of speed-up in time and reduction in space is going much farther. In the germanium *transistor*, a long-lost descendant of the cat's whisker of early wireless days, the

234. After experimental work in the late 1920s by John Logie Baird (1888–1946) a 30-line television service was transmitted by the BBC from 1929 to 1935. By 1937 a regular transmission service using 405 lines was in operation. The photograph shows a BBC television studio at Alexandra Palace, London, in 1939. The television camera (centre) is not so very different in outward appearance from those in use in the 1960s.

movement of electi ons in a crystalline semi-conductor takes the place of their movement in a vacuum. It has already replaced valves for many purposes, especially where small size is important; and other new materials specially devised for even greater sensitivity will probably supplement it. A similar function is carried out by retentive magnetic substances to provide *decision elements* for storing information.

ELECTRONIC COMPUTERS: CYBERNETICS

But it is not so much in the components themselves but in their connexions that the real novelty of modern electronic devices resides. Again, for the purposes of war, it was necessary to make devices which could add and compute as rapidly as was needed to carry out the

235. A modern computer installation – the 'Atlas' computer at the University of London Institute of Computer Science. The photograph shows the operating area with (left) paper tape readers and punches, (at rear) magnetic tape units, (centre) the main control position and (right) card readers and punches.

complicated operations of direction and range-finding and the computation of shell and rocket trajectories. These made it possible towards the end of the war to develop the first fully-electronic computing machines. As computing machines they started where the mechanical computing machine left off more than a hundred years before, when Babbage had attempted, at enormous cost, to set up a machine to calculate mathematical tables more quickly and more accurately than human computers could do. At the moment we are only beginning to sense the possibilities of electronic computation. Here we have a generalized means for translating into movement of electrons the complicated and orderly processes that are carried out in the computer's mind.

Such a machine can not only carry out precisely orders given to it, but it can – and this is the essential novelty – react to the unforeseen situations dependent on the value of the first stages of its own calculation. Like the servo-mechanisms, of which it is a highly specialized and refined type, it can react to contingencies, and even already begins, in selecting concordant and rejecting discordant results, to show some of the characteristics of *judgement* and of *learning*, in finding out easier ways of doing things that have been done once and so to a certain extent making up its own rules as it goes along. In all this it must carry within itself a large number of data or *bits* of information, some provided from outside, others generated by the operation of the machine and requiring to be held for further use, held indefinitely but releasable at call. This is the *memory*, the essential feature of electronic computing. While a certain number of memories are of a static kind, recorded by magnetic marks on tape or wire, or by assemblies of wires and magnetic loops, others depend on recycling the message indefinitely round an electronic circuit. Wiener has shown in his book that *Cybernetics* (or steersmanship) is a new branch of creative science, linking mathematics, electronics, and communications engineering. It is guided by *information theory*[6.123] and has links with the physiology of the nervous system and with psychology itself. The possibility of constructing what are effectively thinking machines, no matter how low the level of thought, is certain to have a profound influence not only on science but on economics and social life (pp. 942 f.).

## THE WAVE NATURE OF THE ELECTRON

While the control of long electro-magnetic waves was providing new telescopes, that of electrons themselves was providing new microscopes. De Broglie in his theory of 1924 had suggested that each electron was accompanied by a wavelength inversely proportional to its velocity. Three years later Davisson (1881–1958) and Germer accidentally discovered the diffraction of electrons by crystals, analogous to the diffraction of X-rays by crystals discovered fourteen years before. This discovery might have been made in an attempt to verify the de Broglie theory. In fact it was hit on purely experimentally, and belatedly at that. The diffraction of electrons might even have been observed before the discovery of X-rays, for thin pencils of electrons had been shot through metal plates as far back as 1894, but no one had thought of photographing the emerging beam. If electron diffraction had been observed and then the wave nature of the electron deduced from it, the whole course of the development of twentieth-century physics would have been altered and probably very much accelerated, though the same discoveries would probably have been made in a different order.

## THE ELECTRON MICROSCOPE

Even before the parallelism between electrons and light in their dual role as particles and waves was recognized, the idea of using deflecting electric and magnetic fields to focus them was beginning to be used. We now know how to concentrate and focus electrons to employ all the techniques of refraction and interference already in use in normal optical instruments. The difficulties of doing this at first were essentially experimental, as electrons can move freely only in a vacuum and the 'lenses' for them had to be immaterial, electrical, and magnetic fields, but they were overcome as techniques improved and a new science of electron optics grew up. Its greatest triumph was the *electron microscope*. The ordinary light microscope is limited in the size of the object it can see by the coarseness of the waves it employs, and although to our senses a wave of light is an extremely small thing – less than one fifty-thousandth of an inch – it is still very large compared with the dimensions of an atom, in fact some 2,000 times as large. Now electron waves can be made much shorter than this, and it is convenient to employ them with a wave-length of about a tenth of an atomic diameter. By a combination of electric or magnetic lenses it should accordingly be possible to imitate a microscope in which the magnification can be made a hundred or a thousand times greater than can be obtained with a light microscope. Ruzcka succeeded in doing this and built the first electron microscope in 1937. Since then they have been greatly improved

236. The electron microscope, using a beam of electrons instead of a beam of light, is capable of magnifications of hundreds of thousands of times. Instruments are shown being assembled at the Shanghai Electron Optics Research Institute.

in range and magnification, so that objects as small as single molecules can be seen distinctly. When the object can itself emit electrons, as can a hot tungsten point in the field emission microscope, it is possible to see all the individual atoms.

The electron microscope is even a greater advance on the ordinary microscope than the microscope was on the unaided eye. It enables us to see and reproduce on photographs the whole range of structures, from those clearly visible in an ordinary microscope down to those of practically atomic dimensions. It is the most direct way of bringing the structure of small objects into the range of our ordinary senses. As such it has a great philosophic importance because it gives a visible reality to unities such as molecules, which were first thought of as abstract hypotheses. Structures of such dimensions are the most interesting and significant for the understanding of the characteristic properties of life (pp. 904 ff.).

### 10.5 Solid State Physics

Though its first appearance must be dated back to the twenties of this century, the importance of solid state physics has been growing with increasing rapidity and it is only in the last few years that it has become dominant in most practical and theoretical aspects of physics. It is necessarily a consequence of the first understanding of the structure of the solid state, both in its regular form as crystals and in its irregular form as glass. Until the thirties the properties of solids, essential in all aspects of engineering progress, had to be either sensed by hand or eye tests or measured by various types of gauges. These measurements were of essential practical value but lacked any pretence to scientific significance.

When science first entered into the study of the solid in the twenties it was to discover that the actual mechanical performance of solids was quite different from what was predicted on the basis of their structure, yet the structure was basically a correct one. This we now see was a structure which belonged to what is called the *ideal crystal*, the crystal in which every atom had its regular place in a three-dimensional lattice. But *real* crystals hardly ever corresponded to this ideal picture. The work of I. Joffe (1880–1960) was the first to show how limited this picture was. By his study of rock salt he was able to show that, if

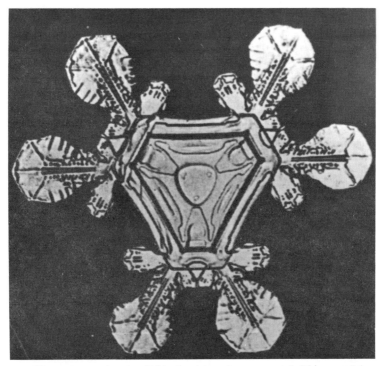

237. Photomicrograph using light of a trigonal snow crystal. This crystal has additional growths at each of its three 'corners'; the lines show how the crystal has grown.

surface imperfections could be removed, the crystals could have a strength far greater than could be measured in untreated crystals, and at the same time they became much more plastic.

Similar results had been found for glass by A. A. Griffith who showed in 1920 that there the weakness was due to very fine cracks which initiated larger ones. These types of imperfections gained a general theory due to the work, in the first place, of G. I. Taylor, who explained the mechanical failures by means of *dislocations* or places where the rows of atoms are out of step with each other. Later F. C. Frank and W. T. Read showed how unit dislocations, themselves only capable of moving a single row of atoms, could multiply and generate trains of dislocations capable of producing large displacements. This mechanism

which was by means of spirals developed by rotation around dislocations was also found to account for the phenomena of crystal growth.
This can now be used practically to grow artificially the remarkable
perfect crystals required for many physical instruments.

## THE NEW METALLURGY

Metallurgy, long a highly developed art, had until the mid century
lacked a scientific base. The properties of metals and alloys could be
measured and modified empirically but could not be predicted in
atomic terms. Now, with the new metallurgy, this is becoming possible
and even necessary to meet the extreme demands of nuclear, space and
rocket engineering. One application is the production of super hard or
super tough materials. A consequence of dislocation theory is that all
materials show their greatest strengths when they are very thin, especially in naturally or artificially produced *whiskers*, whose phenomenal
strength was first shown by Z. Gyulai in 1954.

## PIEZO-ELECTRICITY AND MAGNETOSTRICTION

The mechanical properties of crystals are closely related with electrical
and magnetic effects. In the first case we have the phenomenon discovered by Pierre Curie of piezo-electricity occurring in crystals of
lower symmetry. This has been used in recent times either to detect
small changes of pressure, as in echo sounding, or conversely in the
*transducer* to generate supersonic vibrations which can take the place of
mechanical tools in grinding and boring operations, for example in
dentistry. The changes in length due to magnetization or magneto-
striction can be used for similar purposes.

## SUPER-CONDUCTORS

An effect first discovered by H. Kamerlingh-Onnes (1853–1926) at the
beginning of the century was that some metals and alloys lose their
electrical resistance almost entirely at temperatures near absolute zero.
This is the phenomenon of *super-conductivity*. In the twenties Kapitza
and others found that this effect was destroyed by magnetic fields,
which put an absolute limit on the amount of current that a super-
conductive wire could carry before it was stopped by its own field. This
phenomenon was not understood until a more adequate theory was
worked out by J. Bardeen in 1956. More recent still is the discovery that
certain alloys are remarkably resistant to magnetic field interference.
The way was then open to using almost indefinitely persistent currents
to generate very great magnetic fields without serious power loss. This

may be an essential step to making the magnetic bottles necessary for thermonuclear energy production and also gives great hope of cheap long-range power transmission.

## SEMICONDUCTORS AND TRANSISTORS

Probably the most important aspect of solid state physics today is that based on the electro-magnetic properties of slightly imperfect crystals. This was a by-product of the very simple method of physical purification called 'zone melting' (one which I had accidentally introduced as far back as 1927 but had found no use for). Two crystals of germanium, for instance, can be prepared, one with just a minute excess of electrons and one with a lack of them – or holes. Together they form the sensitive *p–n junction* which is the basis of *transistor* technology.

238. Dislocations in crystals may be detected with the electron microscope. This transmission electron-micrograph of an iron-nitrogen alloy (o·o1 wt. per cent. nitrogen) shows the material after being quenched from 550°C and then aged 3 days at 100°C, after which it was strained. Dislocations (showing as lines) occurred in the alloy's crystalline structure but iron nitrate precipitate has acted as an obstacle to the dislocations. Photograph taken at the National Physical Laboratory, Teddington, Middlesex.

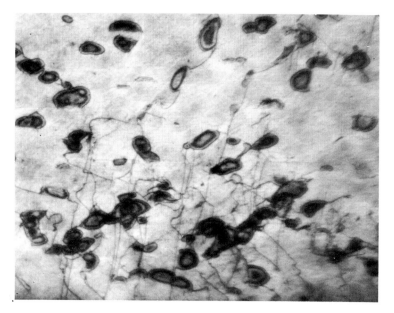

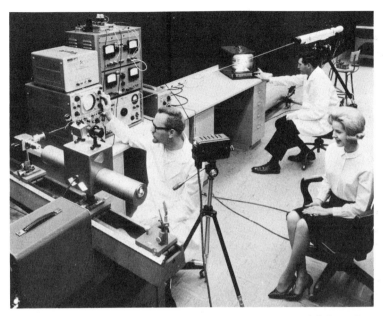

239. A laser in operation with the light beam modulated to transmit information. The young woman (right) is sitting in front of a television camera. The modulated electronic signal from the camera is fed to the laser (left, front) and the modulated light beam then travels to the telescope-shaped receiver (right, rear). The modulated light beam is analysed and applied to the television monitor at the end of the desk (right). It is estimated that it will be possible to carry a thousand million simultaneous telephone conversations or one thousand separate television channels on a single laser beam. Photograph by North American Aviation, Inc.

## LASERS

A radically different mode of light emission from any before known was provided in 1960 in the discovery by T. H. Maiman of the laser principle. This did not in the first place come from optical studies but from that of microwave developments by which molecules could be made to vibrate in resonance with the electrical field. This led to the invention of the first *masers* or Microwave Amplification through Stimulated Emission of Radiation. It was discovered by a similar circuitry that slightly impure crystals such as rubies, when properly cut and coated, can be made to reflect visible light in a *coherent* way with all the atoms contributing in harmony, and hence with enormous intensity, narrow beam and purity. Thus the *laser* principle is the greatest practical physical discovery of recent years. It is far more than a device. The change from incoherent to coherent emission of light is scientifically equivalent to that from a liquid to a crystal, or, in human historic terms, that from an irregular horde of warriors to disciplined soldiers advancing in rank and file.

More recently still it has been found that the laser principle and the transistor principle can be merged together and a simple electric current can be made to generate light at a *p–n junction* and this light, by a suitable adaptation of the size of the crystal can be made to *lase*, that is, to send out enormously intense beams in a specific direction. It is clear that the laser principle is only the beginning of an enormous and wide-ranging development of high-precision physics, absolutely essential for the proper control systems in space and hence, in this military world, not likely to lack money to develop it. The laser is indeed already the death ray, dreamed of by science fiction writers fifty years ago, as by Wells in *The War of the Worlds*, but, fortunately, now it is likely to find a more defensive role as an anti-nuclear rocket device.

Analogous in many ways to the laser effect, but produced by spontaneous radioactive changes, is the Mössbauer effect, for which the discoverer was awarded the Nobel Prize in 1961. Here, a particular isotope of iron Fe–57 was found to emit gamma rays in extraordinarily long wave trains, thus with extraordinarily accurate frequency. If the atoms were in fixed positions, such as they are in a crystal of iron, these trains were emitted coherently, providing an extremely fine standard of frequency, a hundred times better than any known before. They were capable, therefore, of measuring very small movements and even of verifying Einstein's theory in a terrestrial laboratory (pp. 742 ff.).

It is already evident that the solid state contains enormous possibilities for the reinforcement and refinement of the existing phenomena of physics and further that we have only really begun to touch its possibilities. So far the remarkable effects of peculiar substances such as the lodestone, the first magnet, have been discovered in the first place largely by accident. It is evident that much more effective results could be obtained by a systematic search guided by theory. We may expect in this sense to see the bounds of physics enormously expanded with corresponding applications in all the other sciences.

## 10.6 Physics and the Structure of Matter

MOLECULAR ARCHITECTURE AND CHEMISTRY

Long before the electron microscope was developed, a far more power-ful, though indirect, way of seeing even finer structures had been de-veloped, following the original discoveries of von Laue and the Braggs on the diffraction of X-rays by crystals (p. 738). These methods of crystal structure analysis have now been so perfected that it is possible in a very large number of cases to determine the detailed positions, sizes, and shapes of atoms in quite complicated molecules. For example, the structure of the penicillin molecule was first worked out purely by X-ray methods before it was confirmed by chemical analysis.[6.164] Later, Pro-fessor Hodgkin, who was largely responsible for the penicillin structure, led another team on the structure of Vitamin $B_{12}$, the anti-pernicious-anaemia factor. This complicated molecule, with atoms containing long loops as well as rings, was completely analysed with only incidental help from chemistry. This was made possible only by the extensive use of computing machines. It now appears that we have a reliable, but as yet expensive and slow, method for working out many of the key complex molecules of biochemistry and ultimately of the proteins themselves. X-ray analysis showed atoms as definite, more or less spherical, bodies of different sizes according to their inner constitution, and within molecules or in crystals having relatively constant and measurable distances between them. The imaginary pictures of molecules which Kekulé and the nineteenth-century organic chemists had drawn to illustrate the logical consequences of chemical reactions were shown to have a material and spatial base. The X-rays are not the only short-wave-length radiations that can be used to unravel the structure of molecules and crystals. Electron diffraction has also been widely used, especially for studying surface effects, often of vital practical importance, and for determining the structure of molecules in gases. More recently the diffraction by crystals of neutrons from piles has also been used. They have the great advantage of giving information on the nuclei of atoms instead of on the electron clouds. This has revealed the existence of anti-ferromagnetism, in which the magnetic moments of atoms are arranged to neutralize each other instead of supporting each other as they do for instance in ferromagnetic iron.

INTERNAL VIBRATION OF MOLECULES

The picture of molecules given by X-rays was necessarily a static one. It was a long exposure in which any internal movement was blurred;

but twentieth-century physics was to supplement this deficiency as well, and give the information on the dynamic behaviour of molecules equivalent to providing a cinema film of their movements. This was the result of applying the quantum theory of the spectra of molecules, particularly in the infra-red, where the period of vibration of the light could be in tune with the natural vibrations of the atoms in the molecules. Alternatively, as Raman and Mandelstam showed in 1928, the value of these frequencies could be found through the minute changes that occurred in the colour of visible light scattered by molecules. The rates of vibration in different parts of the molecules were to furnish extremely accurate measures of the forces holding the atoms together in these molecules. The new physical methods thus built up a *quantitative* physical picture complete with distances and forces, of what had before been a purely formal account of how molecules held together in terms of such *qualitative* concepts as valency and affinity.

NEW THEORIES OF CHEMISTRY

By 1920, with the theories of Kossel (1888-1956) and of Lewis and Langmuir based on the simple Bohr atom, which could gain or lose electrons to become a positive or negative *ion*, it was possible to reinterpret inorganic chemistry in physical terms. This was an enormous gain in rationality. The chemistry of the nineteenth century could find simple

240. Naturally-occurring crystals of mineral substances often have a great beauty of their own as well as providing the X-ray analyst with a clue to their structure. Crystals of quartz.

formulae for compounds but could not explain either the properties of these compounds or even why some were formed and not others. The new understanding of the atom now made it possible to begin to explain both kinds of facts, and to make chemistry depend less on memory and more on deductions from a few simple principles. The general field of chemistry could now be divided into four sub-fields: that of the rare gases, where all electrons remained attached to atoms; that of metals, where there was an excess of electrons; that of non-metals, where there was a lack of electrons; and that of salts, where exchanges had taken place between the metal and the non-metal ions. This is the modern justification for the Arab-Paracelsan spagyric system of mercury, sulphur, and salt (p. 398). The analogies from external appearance on which it was based find their explanation on quantum theory grounds (p. 619). With the development of the quantum theory this general picture could in turn become quantitative; in the case of salts or ionic crystals the forces holding the whole crystal together could be calculated in terms of known electrostatic potentials.

## THE CHEMISTRY OF MINERALS

This had an immediate effect on the understanding of the complex chemistry of the minerals and rocks. Sir Lawrence Bragg's detailed X-ray analyses, combined with V. M. Goldschmidt's (1888–1947) wide-ranging surveys of all the elements and Pauling's theoretical insight, showed that the stability of mineral structures, and hence their occurrence in the earth, depended on very simple considerations. A stable mineral in fact occurs when appropriate numbers of its constituent atoms, which may be considered as spheres of different sizes, pack snugly and regularly together. The mineral world, from being a chaos, fell into order, and the new knowledge was immediately valuable for understanding the distribution of the elements in the rocks and hence learning where to find them. Crystal structure indeed was to prove the key to the formulation of the principles of *geochemistry*, by which the short- and long-term transformations of the rocks through erosion, deposition, folding, and volcanic action could be followed.

## THE ELECTRONIC THEORY OF METALS AND ALLOYS

An advance of greater practical importance was made by the application of X-ray analysis to metals. These proved to have exceptionally simple crystal structures, which explained the ease with which they alloyed with one another. Here the number of free electrons, which make metals at the same time reflecting and electrically conducting, was seen to have a predominant influence, and a beginning could be made in a rational and not merely trial-and-error metallurgy. The structural studies did more: they explained the primary, economically valuable properties of metals – their plasticity and hardening, the means by which metals can be forged, rolled, and drawn – and made possible the beginning of a rational control of these processes (p. 789).

## THE QUANTUM THEORY OF VALENCY AND INTERATOMIC BONDS

The problem in the case of compounds between non-metals was, however, very much more difficult. It was not till 1927 that the first clue to the nature of the forces between them was found. They were attributed, in a way only understandable in terms of the quantum theory, to the possibility of the exchange of identical electrons shared jointly by a pair of atoms. It was not till 1934 that a quantitative account of a homopolar or electron-sharing bond was worked out by Heitler and London (1900–54) and applied to the simplest case, that of the hydrogen molecule with two protons and electrons. Even though this method could not be applied quantitatively to the more complex cases, it did bring a physical understanding to most of the hitherto entirely arbitrary and merely experimental facts of chemistry. It explained the general nature of chemical reactions, and why in each reaction a certain amount of heat was liberated or absorbed, corresponding to a change in energy levels of the electrons in the initial and final stage. It also threw light on the most important practical developments of twentieth-century chemistry, those of reactions assisted by artificial catalysts or natural enzymes, both of which act by lowering the energy required to bring about the beginning of the chemical reaction, though they do not affect its final state. It also illuminated the mechanism of *chain reactions*, which either in the rapid form of combustion in an engine cylinder, or in polymerizations for making plastics, have become of major industrial importance.

## RELATIONS OF CHEMISTRY AND PHYSICS

It must, however, not be imagined that as a result of all these advances chemistry has become merely a branch of physics. What has happened is that physical theory and physical experimental methods have increasingly interpenetrated and rationalized the old qualitative ideas and the rule-of-thumb practice of earlier chemists. Chemistry has grown in complexity in dealing with more and more intricate and unstable compounds as rapidly as, if not more rapidly than, its central doctrines have been transformed by physics. Physics is a tool to the chemist just as chemistry is a field of intellectual exercise to the physicist.

THE EARTH SCIENCES: GEOLOGY AND GEOPHYSICS

The status of the sciences of the earth – geology, oceanography, and meteorology – is different in kind from that of the basic sciences of physics and chemistry. This is because they lack the same degree of generality, referring as they do to particular places and times rather than legislating for all places and times. They involve more descriptive and historical and less logical and mathematical elements. They are -*graphies* rather than -*logies*. For that reason, though they have enormously increased in range, the changes that have taken place in them are due largely to new techniques and new ideas imported from physics and chemistry.

In the first part of the twentieth century the development of geology was essentially in the collection of new facts. They have, however, been

241. The 'Mohole' project to drill below the sea bed and through the Mohorovičič discontinuity in the Earth's crust to the mantle is a large undertaking. The first stage of the United States attempt in March 1961 penetrated two miles of water, 500 feet of sediment and 50 feet of rock. The drilling rods, which are connected one above the other in the drilling derrick as penetration takes place, are seen on deck.

given enormously greater precision and extent. Under the pressure of an ever-increasing demand for oil, coal, and metals, methods of survey have been completely transformed. A new science of *geophysics* has arisen by which the most refined gravity, seismological, and magnetic measuring implements have been adapted for use in the field or even in some cases from the air. The information they give about the nature of strata thousands of feet down has been correlated with that from test bore-holes.

The greatest of these, the so-called 'moholes', are already well advanced in preparation. They are to be drilled both by United States and Soviet geophysicists in the thinner parts of the crust of the earth, that are about five kilometres deep under the ocean beds, as against the thirty or more which covers the continents. Their object is to reach below the discontinuity found by A. Mohorovičič (1857–1936) by seismic methods, separating the light from the heavy rocks of the crust. The old-time geologist with his little hammer is as much out of place as the old-time prospector with his donkey, pick, and pan. In their place go armies of engineers and scientists, with aeroplanes, trucks, and drill gantries, their work guided by structural theories, their results checked by base laboratories. It is in this field that the new socialist economies, freed from the restrictions and secrecy of rival commercial exploitation of minerals, have gone far ahead; Azerbaijan turns out more native field geologists than England.

In the decades since the war, however, these mass observations had been supplemented by those derived from other methods and as a result a great revolution in geology has taken place, at least as great as that which established the succession of the strata and the difference between sedimentary and igneous processes at the beginning of the nine-teenth century (p. 641). It is now increasingly recognized that the pattern of the present distribution of land and sea, or continent and oceans, on the world is quite different from what it was in the early geological periods. Not only has the whole of the earth apparently shifted but the continents have moved about according to a hypothesis first put forward by Wegener (1880–1930) in 1912 on the grounds of how the continents seem to fit naturally together like bits of a jig-saw puzzle and on the corresponding distribution of animals and plants. This idea was scouted for half a century by the geologists who could not see how such solid things as continents could move, but now the evidence is becoming overwhelming that they do.

This evidence comes from two quite disparate sources: one due to the beginning of the exploration of the geology of the four-fifths of the earth's surface covered by the ocean. The great advances in physical oceanography (p. 800) have revealed an unsuspected pattern at the bottom of the ocean of a system of rifts running down the middle of all the oceans and round the continent of Antarctica. It would appear that these rifts are steadily opening, the process is going on now and oceans like the Atlantic really represent the pushing out of the continents on both sides to be piled up as mountains in the Himalayan, Alpine and Andean systems.

Other evidence comes from the study of paleogeomagnetism. The rocks are found to retain the directions of magnetic force that prevailed at the time they were laid down. It is thus possible to find, for instance, in India that they were once near to the South Pole and that much of Britain was in the arid desert zone near the equator. These studies have, incidentally, proved extremely useful in looking for sources of salt and oil.

It is further recognized that these changes on the surface must correspond to much greater changes in the depth of the earth and in fact are generated by such changes. The picture of the stable earth is being replaced by one with its hot core generating a series of convection cells such as those that occur in hot weather in the air, but, instead of periods of a few hours, they have periods of millions of years and move slowly with immense inertia at speeds of the order of a centimetre a year. The new picture of the interior of the earth is still only being formed; there are disputes of a radical nature between those who believe that the earth is growing and opening up the ocean beds, and those who consider it to be shrinking and piling up the mountains: both may be true at the same time. The whole of the phenomena of mountain building, volcanoes and earthquakes and magnetic changes can be fitted into this general picture of the earth as a dynamical heat engine.

The full scientific value of this mass of new data still requires to be extracted. Combined with geochemistry and supplemented by model experiments along modern engineering lines, the new geological and geophysical data should become the basis for a full quantitative explanation of such phenomena as mountain building, volcanoes, earthquakes, and glacial periods. On the historic side of geology an immense advance has been made in using radioactive changes to measure the absolute age of strata, so that dates are now as indispensable a part of geological as of human history. The use of isotopes in tracking down the precise origin and dates of different formations is only just beginning. It is already taking the place of fossils for dating formation and extending these dates to thousands of millions of years in the virtually fossil-free pre-Cambrian. It has already shown that life is more than half as old as the earth itself – 2,700 million years.[6.220] There are already sufficient indications from the experience of the new methods that a great transformation in the science of geology is well on the way, but it can be completed only when every part of the earth is open to the use of the people who live there, and when the mechanical and scientific abilities of mankind can be used to discover and utilize natural resources for constructive and not destructive ends.

OCEANOGRAPHY

While in the study of the solid crust of the earth it is the structural and historic elements that predominate, in that of the waters and airs it is the dynamic element and the rapidity of change which need to be understood. The classical days of oceanography lie in the nineteenth century, when the charting of ocean currents and the sounding of the deeps were a natural accompaniment of the opening of world-wide trade and the laying of submarine cables. Its development in the early twentieth century was more extensive than spectacular. Data on the physical conditions of the oceans have been steadily accumulating, and

242. Analysing sea water at various depths is part of the technique for a comprehensive study of the oceans. In this photograph, a water-sampling bottle is being attached to an hydrographic wire before being lowered.

the laws of evaporation and of tide and wind-driven currents have been elucidated. The greatest advances have been made on the edges of the ocean basins, the continental shelf furrowed with sinuous and deep canyons of still unknown origin, which have been studied with the anti-submarine device of the First World War – the piezo-electric echo sounder. The coastal landing operations of the Second World War led to the first really quantitative study of beaches and of the waves and currents that serve to form them. Since the war the most exciting study has been that of the deep-sea bottom, which men are beginning to visit in bathyscaphes. The long cores that can now be extracted from the deep-sea oozes correspond to tens of millions of years of slow deposition, and their interpretation gives clues to the climates of earlier ages. Deeper still, explosive echo-sounding has traced the deposits right down to the crystalline crust. Here oceanography marches with geophysics and seismology and the interest in the results is more than academic. Hitherto man has only exploited the riches under the earth; the far greater expanses under the sea still remain to be tapped.

METEOROLOGY

The air, on the other hand, only came fully into its own in the twentieth century, when the need of air travel in peace and even more in war put a high premium on an hour-to-hour knowledge of temperatures and winds. The knowledge required had also to extend upwards, well out of the range of old ground-locked meteorological stations. One of its first fruits was the discovery in 1900 of the upper limit of our disturbed lower air, the troposphere, and the existence of the smooth-flowing empyrean of the stratosphere. The next crucial discovery came in 1918 with Bjerknes' (1862–1951) polar-front theory of cyclones.[6.22] The cyclone itself was hardly a discovery. It is a phenomenon scarcely possible to avoid noticing; the Chinese sky dragon, terrible but ultimately beneficent and rain-bringing, is a personified tornado. The first accurate description of one was given by Dampier in 1687; the first explanation in terms of masses of rising air set twirling by the rotation of the earth was put forward by Espy in 1841.

The crucial idea which Bjerknes added was the concept of separate masses of warm and cold air interacting only on inclined planes of contact – the cold and warm fronts – with the production of clouds and rain. Bjerknes' theory was an indirect or perhaps negative consequence of the First World War. Cut off in Norway from foreign meteorological information, he was forced to think out an independent way of predicting the weather. By introducing a third dimension

243. The meteorological balloon, carrying radio transmitter and air sampling equipment, is launched regularly by many nations to assist in weather forecasting and climatological analysis. Balloons are released from land and sea stations.

into meteorology Bjerknes anticipated the enormous new importance of the physics of the upper air that was to come from the urgent needs of aviation. In the Second World War this need was partly met by the use of radio aids, notably the radio sonde, which broadcasts meteorological information from accurately localizable balloons, and by the direct use of radar, particularly valuable in the study of storm conditions. Even steady rain has a radar-detectable flat ceiling where it is formed from melting snow. The overall range of meteorological information, both in height and geographical extent, has been immensely increased through the use of weather satellites which can, for instance, cover the whole range of cyclonic disturbances covering thousands of miles in a single photograph. Despite all this new wealth of information, and even despite the electronic computing machines which are begging to be used to reduce it to manageable dimensions, meteorology has still to become a full science with quantitative laws linked with the rest of physics.

A new aspect of meteorology was revealed in 1946 by the use of crystal seeding for promoting rain formation in clouds. Though the method is still far from reliable, it marks the first intentional human intervention with the weather. Bowen claims that meteor dust trails may also have this effect and thus bring about rain all over the world round certain days of the year.[6.28] If this is so the condensation produced by the fission-bomb products may, besides producing radioactive rain, also represent a large-scale human unintentional interference with the weather.

244. The 'seeding' of clouds with crystals can cause rainfall. The effect of cloud seeding as observed from an aircraft.

245. The Model 'T' Ford, 1923. Mass-production was used and from the manufacture of the first model 'T' in 1908 until production ceased in 1927, a total of 15 million were sold.

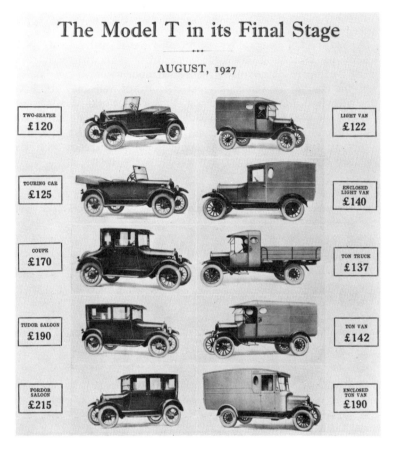

# The Model T in its Final Stage

### AUGUST, 1927

| | |
|---|---|
| **TWO-SEATER** £120 | **LIGHT VAN** £122 |
| **TOURING CAR** £125 | **ENCLOSED LIGHT VAN** £140 |
| **COUPE** £170 | **TON TRUCK** £137 |
| **TUDOR SALOON** £190 | **TON VAN** £142 |
| **FORDOR SALOON** £215 | **ENCLOSED TON VAN** £190 |

## 10.7 Twentieth-century Technology: Engineering

We have now outlined the progress and the interrelations of the physical sciences in the first half of the twentieth century. It remains to follow the effects of these developments on the general technique and industry of the period. The difficulty here is not, as it was in previous centuries, to trace the connexions between science and industry, but to be able to treat them separately, even for descriptive purposes. This has already been exemplified by the need to describe the radio industry as an integral part of the advance of physics. It is apparent that the effect of science on industry is both more rapid and more far-reaching than ever before, and indeed in the Second World War and post-war periods science is quickly becoming an integral and inseparable part of industry. From the beginning of the century, in some industries such as the chemical and electrical, it could fairly be claimed that science had now more to contribute to industry than to learn from it. By the middle of the century that point was being reached even in the most traditional industries such as agriculture and building.

The development of industry in the twentieth century, though it follows on continuously from that of the nineteenth, has proceeded so fast and so far as to turn the whole process of production into something virtually new. The essential change in the first half of the century was in the methods of production from machine-aided craftsmanship to *mass-production*, and this is likely to give way in turn in the second half to *automatically controlled production* where new mechanisms, largely electronic, take the place of the unskilled operators of the semi-automatic machines of today. With this change in the productive methods of single articles goes a much greater interlinking of different industries and the turning of traditional and craft occupations, such as agriculture and building, into mechanized industries. The following sections will contain an outline of the chief developments in mechanical and chemical industry and the parts played in them by science. The electrical industry has already largely been covered in discussing the developments of physics, with which it is inextricably connected.

MASS-PRODUCTION

Mass-production is essentially an organizational rather than a technical innovation. Its elements, interchangeable parts and the assembly line, have existed since the late eighteenth century (p. 595). Characteristically it was Eli Whitney's gun factory in the American War of Independence that first showed that it was possible to make a complicated weapon, not by shaping the different parts to fit each other, but by putting together any selection of them made nearly enough alike to be assembled.[6.74; 6.107] The means for doing a series of operations rapidly one after the other had also been worked out practically in the slaughter-houses of Cincinnati about 1870 by the use of the overhead assembly line.[6.74] The linking of these two came about, however, only in the first decade of the twentieth century, because it was only by then that it was possible, and possible only in America, to find an adequate market for large numbers of complicated machines – if they could be made cheaply. At the same time this key development required a shortage of skilled and an abundance of unskilled labour, and a minimum interference from the vested interests of an old and heavily capitalized industry such as those in Britain. Such a combination of conditions could only be found in the United States at the beginning of the century, when the farm lands had been fully occupied but were in need of machines and transport, and millions of fresh immigrants were pouring in from Europe.

THE INTERNAL-COMBUSTION ENGINE AND THE MOTOR-CAR

The machine that more than any other was to transform both industry and conditions of life in the twentieth century was the *internal-combustion engine*. Yet its development did not follow directly from that of steam-power plant. This was largely because the pioneers of power and transport engineering in the mid-nineteenth century were too successful, at least in their country of origin, England. The monopoly of stationary steam-engines for factories, of locomotives for the railroad and of marine engines for steamships, held up development of other forms of power, both electrical and internal-combustion, in England.[6.14] Indeed we might have had the internal-combustion engine some thirty years earlier had it not been for the deliberate restrictions imposed by the railways on any kind of road transport. The notorious Bill which demanded that a man with a 'Red Flag' preceded every motor vehicle was only repealed in 1896, and internal-combustion engines had to be developed in countries such as France and Germany, which lacked anything like the engineering experience that could be had in England.

The internal-combustion engine was, though less directly so than the original steam-engine, a fruit of the application of science, in this case that of thermodynamics. The fundamental idea of exploding a precompressed mixture of air and combustible vapour in order to achieve thermodynamic efficiency was due to the French engineer de Rochas (1815–91) as far back as 1862, but it was a long step from that to a workable engine, and many essential details – ignition methods, valve operation – not required in steam-engines had to be worked out. The practical pioneers Lenoir (1822–1900) and Otto (1832–91), who devised the still almost universal four-stroke cycle, and Diesel (1858–1913), who added compression ignition, were able to make efficient engines, but during the nineteenth century their use was limited to a relatively few static gas and oil-engines. Their use for road locomotives or automobiles grew but slowly in the last decades of the century, and even then they were largely built to order for luxury or racing. Henry Ford (1863–1947) started as a back-yard amateur car builder, and rapidly became the most successful of the manufacturers of the new car, because he realized that what was really wanted was a cheap car in enormous numbers.[6.71] This necessitated some measure of mass-production, and at the same time gave an enormous impetus to its further development. From then on all the classical methods of

246. Modern mass-production methods are similar in principle to those used by Henry Ford. A view in a British Motor Corporation factory.

engineering had to be redesigned to enable them to produce identical parts in quantity and without the individual attention of the skilled craftsman.

## THE MOTOR INDUSTRY

Once the cheap car was available, the enormous latent demand hitherto unrealized for individual, family, and goods transport on the roads gave rise to a whole new industry. This should serve as an example of the lack of knowledge of the capitalist entrepreneur of where profits could be found. There is no way of assessing the real need for a new product unless a sufficient number of prototypes is available. But to supply these requires investment in plant, and the difficulty under capitalism has always been to finance such early stages. The result is that the great delay between first invention and first effective use is largely due to these purely financial considerations.

The essential problem under capitalism (already alluded to on pp. 612 f.) of financing early stages of inventions depends on the expected return on investment. Even with as slow a return on capital as 3 per cent, it will hardly pay to put up money for anything that has a reasonable chance of bringing in money only after thirty years. Even then the return would have to be at least ten times the original investment to make it worthwhile. If the development is not a certainty the prospect of finding backers in the early stages is even less. Only quick-return prospects are really worthwhile, and except in fields like antibiotics rarely involve any radically new principle. With money doled out slowly the technical difficulties tend to hold up the new developments, and hence to make them still less profitable. For the ultimate profit has to be made quickly even for patented inventions, for afterwards it goes to the cautious investors, who will put their money only on proved successes. As a result, even in the twentieth century, the average time between the essential idea and the commercial pay-off remains about a generation. Whittle had the idea of the jet engine in the early thirties. It was developed slowly for lack of funds and, despite the military need, was hardly ready by the end of the war. The situation would be different if a very large investor, which in these days could be only the State, intervened. By putting more money in at the beginning, even though most of it would be lost on failures, the development time on the remainder would be cut so short that it would pay for everything even at high rates of interest. As socialist governments get into their stride they will win the race for industrial advance unless capitalist governments change their habits or even their natures to catch up with them.

Once the profitability of motor manufacture was proved capital flowed in readily enough. A new industry grew up which was in a few years to outstrip the older engineering industries, and in large measure to absorb them. The automobile industry was, from the moment of its popular success, highly concentrated, for only the very largest concerns could meet the market demand. Alongside the new chemical and electrical combines the automobile industry took its place at the very centre of monopoly capitalism. It is interesting, but not very surprising, to note that the first large-scale development of the motor-car came practically at the end of the development of the internal-combustion engine, for, with minor modifications in performance, of an essentially technical nature, it still remains what it was in 1880. What is radically new is not the car itself, however its appearance may have changed, but the mass-production methods of manufacturing it, to which we shall return later. The further technical development of the internal-combustion engine into the internal-combustion turbine was to come from another quarter, that of aviation.

AVIATION

To be able to fly like a bird has been one of the perennial dreams of mankind, as shown by the widespread legends of flying men or flying machines and by the early attempts in all lands to imitate the birds. It has especially appealed to scientists, to such varied characters as Leonardo da Vinci (pp. 395 f.), John Damian (*c.* 1500) the alchemist of James IV of Scotland,[6.118] the mathematician Cayley (1821–95), and the experimental physicist Langley (1834–1906). We know now that none of them could have succeeded, at least in sustained flight, for lack of a light power unit; though they would have been able to make and fly gliders as well as anyone can today. Actually, though the scientists pointed the way, and though Langley made a large steam-powered model that flew about half a mile, it was not for the scientists to make the final successful efforts. The problems of flight are so complex that they could not have been solved by the science of the last century; indeed many important ones are beyond the science of today.

The development of flight in actual practice was to be a technical rather than a scientific achievement, resembling the development of the canoe into the ship. There is, however, this significant difference: that whereas the former took something like 2,000 or 3,000 years and proceeded by almost imperceptible steps, the latter was effectively accomplished in less than twenty, and involved literally one decisive jump off the ground. The difference is due to the more conscious and more dynamic technical and social background and tempo of the twentieth century. The early attempts at flight were, and had to be, entirely of an amateur kind. Only enthusiasts would risk the certain financial loss and the serious danger to life and limb that early flying experiments entailed. Lilienthal, the greatest and most scientific of the pioneers, was killed in his glider in 1896. But there were enough amateurs, and the experience that they gained was passed on from one to the other until at last success was achieved.

For sustained flight everything depended on having a sufficiently light source of power, and such a source of power could only be available in the twentieth century from the development of the internal-combustion engine. The Wright brothers, cycle mechanics by trade and aeronauts by inclination, mounted a home-made engine in a plane and modified it until it flew for the first time in 1903. *C'est le premier pas qui coûte.* Once Orville Wright (1871–1948) had lifted his plane off the

247. The first successful powered flight took place on 17 December 1903 near Kitty Hawk in North Carolina. Orville Wright (prone on the lower wing) is at the controls, while Wilbur Wright, running alongside to balance the machine, has just released his hold on the right wing.

ground and kept it in the air for a few feet, the future of aviation was secure. No matter how many accidents, no matter what financial losses were incurred, it was now known that men could fly. Progress in every direction, though still for a decade of an amateur kind, was rapid, simply because there was now available in the new motor industry a range of interest and technical ability able to flow at once into any radically new opening. The immediate profitability of aviation was not apparent, but its publicity value was enormous and could be exploited by the cheap Press. Unfortunately there was only too soon to be an overwhelming demand for the new flying machines. Within eleven years of the first flight the first plane was in battle. From then on the needs of war were to provide a perennial incentive for the development of flight and one that to this day absolutely dominates aviation.

## AERODYNAMICS

Essentially because of its empirical origins, the aeroplane was in its first decades to contribute more to science than it drew from it. It gave rise to the first serious study of aerodynamics, which was to have wide repercussions in engineering and even in the sciences of meteorology and astrophysics. Earlier efforts such as those of Magnus (1802–70) were concerned with the flight of shells. The study of streamlined motion and turbulence undertaken in relation to the development of early aeroplanes was to find immediate application in ship design and in all problems of air flow, from those of blast furnaces to domestic ventilation.

A comparison of the development of the aeroplane in the twentieth century with that of the steam locomotive in the nineteenth shows the enormous effect of the economic and political conditions of the age of imperialism. Even today the locomotive is economically a far better proposition than the aeroplane. Now the locomotive was developed in a period of profound peace and for purely commercial and profit reasons (p. 547). It required big capital, but it could be counted on until quite recently to pay its way. The aeroplane, however, was almost from the start under the wing of the State with an eye always to its war potential. For the thirty years between 1930 and 1960, aeroplane construction was primarily military. In the period immediately after the Second World War more than ninety per cent of such production was of the military kind. Consequently, the technical development of planes was not dictated by their convenience but by their potential bombing or fighting capacity. This ensured that aeroplane construction should develop quickly but it disorted that development away from the service of normal transport and represented in all a fantastic waste of human effort.

JET PLANES

The evolution of the propeller-driven aeroplane had been a straight-line one, from the Wright biplane up to the Super-Fortress, but the demand for ever higher speeds for military purposes has at last broken through the typical conservatism of designers and produced the gas turbine, which has made possible the jet plane. It was characteristic that this development, both in England and Germany, had been seen to be inevitable for many years, yet little encouragement was given to the pioneers and even in the Second World War they arrived too late to be of military value (p. 808).

The subsequent speed of development of the jet plane for military use is one of the characteristics of cold-war application. Intended to supersede, and indeed superseding, all old piston-engined aircraft for fighters and bombers, their career has ended without their being actually used in any major conflict. Apart from the few used in minor operations against ill-armed enemies, they were just made by the tens of thousands at vast expense and sent to the scrap heap. They have, however, completely replaced any other form of plane for civil transport and are working up to speeds well above those of sound even for this purpose.

248. The modern jet aircraft is a large and expensive machine to produce, yet it has completely replaced any other form of air transport for long journeys on the arge air lines. A Super V.C. 10.

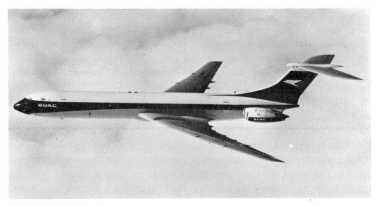

The world, however, is not really large enough for supersonic jet planes. Already the time to reach the airport, go through the formalities, etc., is almost as great as that taken to fly across an ocean or a continent. Nevertheless, air transport for use and convenience, with the planes adapted to their function for short and long distances, is clearly on the way, only held back by considerations of prestige. Now that the military aspect of atmospheric flight is rapidly disappearing, it is possible that cheap, convenient and safe air flight will take the place of all passenger and much freight transport except for short distances.

ROCKETS

The great development, which has taken man for the first time outside the bounds of earth and air to which he seemed to be fixed by his very nature far more than he was fixed to the ground – for he had always the birds to look at – was due not to any striking new scientific discovery but to the use of a long-tried method, a Chinese invention, preceding even the cannon – the rocket. The development of the modern rocket has a curious history. Beginning with ornamental fireworks and sporadic use as a secondary military weapon for incendiary use throughout the Middle Ages, it was first taken up in the 1920s with the deliberate intention of navigating into space, by a number of enthusiasts, considered hopeless cranks, in various countries, particularly in Germany. I was myself among those who speculated on the subject of space colonization and published a book on it in 1929.[6.18]

How long this development might have taken is anyone's guess, but the war provided, fortunately only in its latter stages, a new and what proved a decisive impulse to rocket development. Von Oberth's work was taken up enthusiastically by Hitler, always in search for strange and unproved contrivances. Success came too late and, although the $V_2$ rocket had considerable nuisance value in its attacks on London, these could be countered by the most elementary methods of attacking its bases. In fact, as long as we were limited to conventional explosives with a range of action measured in a few yards, long-range rockets clearly had no military future. The cost of launching, the difficulties of aiming were much too great.

Once, however, atomic weapons arrived this situation completely changed. With the enormous quantity and extent of damage caused by even the relatively primitive fission bombs, it would clearly pay to introduce long-range and even inter-continental ballistic missiles which could afford all the refinements of electronic guidance systems and research into better propellents. The longer the range the more

249. The development of the modern space rocket began, curiously enough, with the vogue for ornamental fireworks in the seventeenth century. Fireworks on water with an imitation of a naval combat, after a copperplate engraving from the *Pyrotechnie* by Hanzelet le Lorrain, Pont-à-Mousson, 1630.

important it is to achieve an hitherto inconceivable accuracy of aim, a mile in several thousand miles. The development of shock-proof transistors instead of fragile valves made this much easier to achieve. The intensive development of rockets was accordingly taken up, in the first place by the Soviet Union and later by the United States. The British started last and never arrived: all their rockets became obsolete before they could even be used. No other country has effectively built any independent rocket system, although there is a danger, following the example of France, that rockets will be considered an obligatory prestige part of all modern armament systems even when the know-how has to be borrowed entirely from outside. With the development of the hydrogen bomb the long-range rocket became the ultimate weapon, which has produced the present horrible situation in which we all await the total destruction of civilization and possibly the wiping out of mankind.

## THE CONQUEST OF SPACE

In the resulting military stalemate, other and less dangerous uses for rockets had to be sought. It was as recently as 1957 that man's conquest of space began with the launching of the first artificial earth satellite in the USSR. This was a first decisive break-out from the earth's gravitational field. But unlike the amateur effort that began powered flight fifty-four years before, it was a complex combined scientific and engineering effort, drawing on the whole of the technological expertise of the two most powerful countries of the world, the USSR and the USA. In launching the sputnik and putting it into a stable orbit the whole world was indeed shown that technically and scientifically the USSR was in the forefront.

It was also the beginning of a full realization that the future, not only militarily but technically, belonged to the development of science. The sputnik's effect on stimulating American scientific education was perhaps its most remarkable by-product. There followed immediately a new kind of technical race in space. More money has, indeed, been spent on the advance of space research in the two countries which control it between them, the USA and the USSR, than on the whole of science in all previous times, eclipsing even that spent on the early development of the atomic bomb. It is estimated that between 1959 and 1962 it was some $5 billion in the United States and, to judge from their success, it could hardly have been less in the Soviet Union.

It would be encouraging to think that all this is for the glory and advance of science, as, indeed, it may turn out to be in the end, once we can secure the common development of all types of space vehicles and the elimination of all rocket-carrying weapons. Space research is already becoming one of the most adventurous and the most technically advanced outlets for human enterprise. There can, however, be little doubt that the same effort of human manpower and brain power could have afforded very much greater advantages at this particular time in the normal development of science which is just reaching a most exciting phase.

Elsewhere I have discussed (p. 813) the scientific results of the first four years of the space age, the landing of an unmanned satellite on the moon and the journeys of the spacemen around the earth preparatory to flights further afield. Until a much later period when the colonization of space becomes a possibility rather than a dream, it is unlikely, however, that the development of rockets and satellites is going to be of much human value apart from weather surveys (pp. 801 ff.) and telecommunications.*

250. The most recent peacetime uses of rocketry have involved the launching of sputniks, lunar and planetary probes, and even man into space. The successful Earth orbits completed by Yuri Gagarin in April 1961 were the beginning of an era of space travel that will doubtless see men landing on the Moon, if not further afield within our Solar System.

## TRENDS IN SPACE-AGE TECHNOLOGY: SPEED

The trend of aeroplane and rocket design is one example of a general tendency, apparent in all modern technique, towards greater and greater speeds. Speed carries with it both advantages and compensating disadvantages. High-speed engines have the advantage of being able to pack more power into a smaller space and, by operating more quickly, to get far more work done or goods transported in a given time. At first sight they would appear to be capital-saving – the vast beam engines of the eighteenth century gave a miserable four to ten horse-power; a thousand-horse-power motor could now be put into one of their cylinders. But this contrast is somewhat deceptive; what is gained in compactness is lost in high working and maintenance costs. We no longer expect the hundred years of continuous service which some of the old beam engines actually gave. As higher speeds require far greater perfection in both materials and manufacture, the original cost also goes up with the speed. These disadvantages disappear where speed and compactness are everything and cost is nothing – that is in war. Even for peace-time uses speed may be economical where it is linked with higher working temperatures, which lead to greater thermal efficiency. In the analogous case of electricity, higher voltages raise insulation problems but involve proportionately smaller current losses, and hence make possible long-range transmission of power (p. 616).

## AUTOMATION

The logical sequel to the development of mass-production with production lines each containing more or less complex machines and linked simply by operatives handling the parts from one to the other, is that of the *automation* of the whole process; that is, the combination of completely automatic individual machines and some form of *transfer* machine for passing the pieces from one machine to the next. In the highly mechanized industries, particularly the motor and engine industry, this process has already gone far. It has been made possible now, on quite a different level of achievement, by the introduction of electronic computers as essential controllers of the whole process. It is no longer a question of combining machines into a single production line now automated, but of combining automated production lines into a total production process, beginning, for example, with the raw metal and ending with the packed machine, tested and labelled.

We are now clearly in a period of transition to complete automation. In the capitalist countries this is coming in piecemeal on a strictly

251. The increasing use of automation is a logical development of mass production. Instructions can be provided by computer as in the case of a slabbing mill at the Spencer Works of Richard Thomas and Baldwin where the operator receives information from an indicator (top right) controlled by a production scheduling computer. The difference between this steel works and one in the last century may be seen by referring to plate 178 in Volume 2.

business basis, that is, where the greatest profit can be made by introducing automation. In socialist countries it is being introduced in a planned way, aimed at a coherent and interrelated set of automated factories. In any case it is the logical next step in technical development and its advent is already having profound economic and social consequences.

COSTS AND CAPITAL SAVING

One by-product of automation is the building up of large single plants with heavy equipment, very expensive and only economic if working more or less continuously. The advent of these plants has led to an inverse relationship of cost to rate of production. The larger the production of objects which are not easily stored, the greater the tendency to put emphasis on sales rather than on production itself, if only to keep the production lines clear. One purely technical solution to this problem is to make the automated machinery simpler, cheaper and more versatile. This is a tendency clearly visible in the electronics industry itself, namely, the idea of breaking down complex machines into unit machines which can be assembled in different ways and so allow for a more or less continuous improvement of the production methods of the whole plant. This process, which has been called *digitalization* of industry, analogous with the operation of the computers themselves, is clearly a development of the future.

At the moment production faces enormous difficulties merely from its rapid rate of advance. Production methods risk being permanently dislocated by the role of obsolescence which is already comparable with the time required to evolve them. In the military field it has long been known that many planes and rockets are obsolete before the design is finished. Production techniques are necessarily always falling over each other. Further, the cure of this can only be found by taking a hair of the dog that bit them, namely, by automating automation, considering production as part of a process involving also the ordered production of production methods and using computers and operational research programming for this purpose.

The effects of automation both industrially and socially will be discussed later (pp. 854 f.).

SCIENCE, COSTS AND CAPITAL SAVING

The pursuit of speed and automation has certainly stimulated science and technology, since the greater the speed the greater the need to understand the processes and materials involved and to raise the standard of specification and workmanship. These are not the only factors driving the engineering industry in that direction. Everywhere economic conditions press towards achieving lower production costs. Not only have things to be done better, but also quicker and with fewer people. High wages secured by unrelenting union pressure promote labour-saving. All this puts a premium on the use of ingenuity and science. There is plenty of scope for both.

ENGINEER AND SCIENTIST

Manufacturing processes, as such, have in the past had precious little science in them. They have grown up by steady and almost imperceptible changes out of the man-and-boy workshop of the early metal age (p. 547). It is not till the twentieth century that any serious attempt has been made to study them rationally and scientifically. Such an attempt implies, among other things, a new relation between engineering and science. In one way it is a return to the situation that existed before the beginning of the Industrial Revolution. In the nineteenth century, with the rapid growth of machinery, an increasing separation had grown between the relatively small number of investigators of new things – the scientists, and the great number of developers and users of these scientific discoveries – the engineers. We are beginning to realize now that it is impossible to have good engineers who are not also scientific, that is, who are not capable of using the techniques of science to analyse and find out what they are doing and should be doing, rather than applying sound experience, common sense, and formulae taken from textbooks (p. 42).

But before the engineer can become a scientist, the scientist has to learn to be an engineer. The weakness heretofore has been that the scientist, in his desire to achieve a solution which is mathematically and experimentally manageable, has deliberately ignored most of the variables which the engineer cannot avoid dealing with: the practical limitations of time and space and of the quality of materials available, and perhaps even more pertinently, because they lie farther outside the ken of pure science, the economic questions of cost and the political problems of management and ownership. Now the fact that these also enter into consideration in every real problem does not make that problem less scientific. It only emphasizes that science has as yet not taken on its full task. It is perfectly feasible to introduce cost factors as variables, both in the production process itself and in the ways of changing that process, and to do so on a quantitative basis with a view to the greatest efficiency. Such calculations were in fact carried out quite successfully in capitalist countries during the war, where the problem could be seen as one of determining how to achieve the maximum effective production for the minimum of manpower and material resources. The problems of organization of industry, which are essentially political and social, though still within the range of science, have, however, a wider scope than the natural sciences, and will be discussed in their place in the section on social science (pp. 1143 f.).

SOCIAL EFFECTS OF MASS-PRODUCTION

The economic and social effect of the growth of mass-production industries was felt mostly in transport and light industry. Once motor vehicles, particularly cars and light lorries, were available in large numbers the process that began in the railway age was completed, and

252. The advent of mass-production of vehicles and multiple building techniques has had many effects of a social kind. The ribbon development of suburban dwellings at Selsdon, Surrey, is a typical consequence.

the countryside as well as the towns was made accessible to goods and passengers. This had immediate economic consequences on the market but even greater social ones in spreading the towns into the country and turning most industrial regions into vast suburbs. At the same time the use of mass-produced agricultural machinery, especially the tractor and combine, has drastically lowered the need for large numbers of women and children on the land. This has helped to break down local particularism and must needs have a levelling effect, increasing always with time, not only locally but between countries and even continents. It does not necessarily lead to greater international understanding, but it tends to turn national into class issues. The awakening of Asia and Africa is facilitated by the introduction of the bus and the bicycle.

Once mass-production was well established in the motor industry it tended to spread to other industries, notably to the new electrical industry. It also accelerated the process of converting the minor textile and food industries, previously carried out in the home, into large-scale industries providing the market with standardized and packeted consumer goods. The mere concentration of such processes into factories in itself gives rise to scientific problems of *quality control* and the adaptation of small-scale techniques to large-scale production. Thus new fields of scientific study concerned with the properties of materials such as *plasticity* and *rheology*, or the science of flow of such substances as pitch or concrete, and with the regulation of processes, are brought into activity. The new sciences contributed in turn to the rationalization of techniques quite outside the fields that gave rise to them. By the mid-century at least a tincture of science was given to all the traditional industries even in their last stronghold, the domestic kitchen.

### BUILDING: CONCRETE AND PREFABRICATION

Only less spectacular than mass-production has been the twentieth-century advance in permanent construction, due to the increasingly intelligent use of steel and concrete. In itself the use of steel is less revolutionary; the steel-framed skyscraper is only the medieval building on a larger scale, and represents in any case a fantastic waste of steel. Far more significant was the introduction of *reinforced concrete* by Monier (1823–1906) as far back as 1868, but only coming into its own in the 1920s. Here a rational combination is sought between the mass

253. The social effects may expand still further with the development of prefabricated buildings on a large scale. A prefabricated 'Instant Schoolhouse'; the building is based on four-arm columns with prefabricated walls attached. The building, by the Massachusetts Institute of Technology, may be extended at will.

and comprehensive strength of the concrete and the tensile strength of the steel. The further step taken by Freysinnet (1879-1962) in 1928 was to put the steel under tension, and thus produce, in *pre-stressed concrete*, a material hardly inferior to steel in lightness and resilience. The use of reinforced concrete has vastly increased the dimensions of man's structures in relation to Nature, as buildings, roads, and dams bear witness. Combined with heavy mechanical excavating and dredging plant, it has effectively given man the power to alter on an ever-increasing scale unfavourable geographical features, to divert rivers, and cut through mountains (pp. 965 f.). At the same time a long-delayed revolution is taking place in the age-old tradition of building, where instead of building up brick by brick and finishing by hand on the spot, more and more units of building are *prefabricated* and building itself becomes essentially a mechanically assisted assembly process. This development has been slow and still faces immense resistance, but it is coming at last, under the pressure of the need for convenient and cheap dwelling space. The problem is, however, more than a technical one. Houses are part of mankind's whole pattern of living, and to reconcile tradition and efficiency will call on the highest abilities of architects and engineers. The enormous growth in the size of cities, whose population has doubled in the last fifty years, puts a premium on this combination of the artist and the scientist. Problems of planning and building are wider, however, than those of any existing discipline. They call for a new combined discipline, involving sociological, biological, chemical, and mechanical studies, with much experimentation on different scales and careful observation to find the solutions to the problems, which, if left to themselves, will choke mankind with his own productivity.

## 10.8 Chemical Industry

The chemical industry is second only to the electrical industry in the degree to which it has been transformed by science in the present century. As a result it has become the central industry of modern civilization, tending, because of its control over materials, to spread into, and ultimately incorporate, older industries such as mining, smelting, oil-refining, textiles, rubber, building, and even, through its concern with fertilizers and food processing, agriculture itself.

The use of the chemical industry for making fertilizers is already, and will be even more so in the future, a major factor in increasing the world's food supply to deal with an increasing population. Such fertilizers are not only those supplying nutrients like nitrates and phosphates, but those using special polymers which stabilize soil structure and turn poor into good soil. With abundant energy the amount of fertilizers of both kinds can be increased till all soils acquire the existing fertility of such soils as those of Britain and Denmark today.

The entry not only of chemistry but to an increasing degree also of physics into the chemical industry has here led to a radical break with the dirty, back-yard chemical industry of the early nineteenth century. Reliance on mere modification and increase in scale of traditional chemical operations is giving way to a consciously designed chemical plant, applying in a calculated way the results of the laboratory to full-scale operations. Such operations require a control quite different from that of the old chemists, one depending on the use of instruments rather than experience and rule-of-thumb methods. It has created the new profession of the *chemical engineer*, while the physical chemist and ultimately the physicist are also coming to play a direct part in chemical industry.

254. The continuous-flow chemical plant is to be found in many aspects of chemical industry, and particularly in oil and petrochemicals. The modern large refinery, with its associated chemical manufacturing plant, covers a large area. The Pernis refinery in the Netherlands occupies 1,235 acres and has a waterfront of 4 miles.

CONTINUOUS-FLOW METHODS, CATALYSIS,
AND THE SYNTHETIC APPROACH

The two great features that distinguish twentieth-century from nine-teenth-century chemical practice are the use of *continuous-flow methods* and of *catalysts*. The use of continuous-flow processes instead of batches was the chemical equivalent of the assembly line, and indeed long preceded it. This involves a much more complete control of every stage, and consequently increases the importance of the use of physical methods of instrumentation and automatic control. The use of elec-tronic computers in the chemical industry has gone further than in any other and the consequent reduction of human interference has led to the achievement of a far greater precision in the control of production rate and quality. It is now being recognized that what might be called a generalized chemical or continuous flow rather than batch process is adaptable in industries which were previously manually or machine-operated, particularly in such industries as metal production, smelting and even fabrication. The introduction of such chemical processes as oxygen and low-temperature reduction, for instance, for iron has brought the basic heavy industries into the realm of chemical industry. For the moment only tradition and ownership arrangements prevent them being totally assimilated. A flow line, starting with the crude ore and ending with a continuous flow of cold steel strip, is essentially a chemical operation. The other great development, which is only now beginning to show its full power, has been the use of catalysis on a mass scale. Catalytic processes are old in chemistry, but the modern use of catalysis, especially in relation to oil and gas chemistry, is essentially so different in scale as to constitute a new era in chemistry. Purification and modification of chemicals are giving way to radical *syntheses*.

In the past chemical products have been made from natural products by a process of separation and transformation. In an extreme case like that of coal, a highly complex natural product is broken down step by step by distillation and then sorted out into its many by-products, which in turn may be transformed into more valuable chemicals. In contrast to this, modern practice, starting with the same or similar materials, makes no attempt to separate existing compounds, but breaks everything down to the simplest compounds or even the ele-ments – the new universal materials of chemistry are diatomic mole-cules such as hydrogen, carbon monoxide, oxygen, and nitrogen. From them, by the use of catalysts, are made all the old and new pro-ducts of chemistry, especially the products formerly obtained from

Nature, but now required in larger quantities and in greater purity than Nature will provide, such as high-performance fuels, artificial rubbers, and the great variety of plastics and fibres.

### POLYMERS AND PLASTICS

All these substances, except the fuels of low molecular weight, are what are called *polymers*, necklaces of molecules automatically strung together by a chain reaction, usually induced by a catalyst. In a polymerization *chain reaction*, in contrast to the violently dissipative chain reactions of combustion or nuclear fission, each new section added to the molecule makes a further addition possible. If molecules are added in one dimension the result is a fibre, if in many branched chains a resin or so-called plastic. The unravelling of the mechanism of chain

255. With the production of petrochemicals and the development from them of plastics, many natural materials have been replaced. Moreover, new materials are often more suitable or lend themselves to mass production more readily than those previously used. Plastic skeletons are now being widely used in medical schools, hospitals and universities.

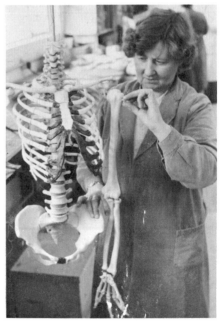

reactions and of polymerization by such chemists as Semyonov and Melville marks one of the most significant chemical advances of the century.

New methods of polymerization, using solid catalysts, seem to depend on another mechanism, the new molecules apparently being stripped or zipped off the catalyst. This production of so-called *isotactic* polymers involves lower pressures and temperatures than with chain reaction and produces polymers of more regular and better quality. Synthetic rubber produced in this way is actually better than the natural variety. Since the war a further step has been taken, through the work of Ziegler and Natta, who by the use of solid catalysts have been able to produce polymers with regular rather than chance arrangement of the chains, thus producing materials of much greater strength and uniformity. These chemical advances are in part the cause and in part the consequence of the development of the new aspect of the chemical industry, that of *artificial fibres*. Many of them, like nylon and polythene, are now household words, though thirty years ago they were not even thought of. In the formation and processing of polymers the new rational chemistry, aided by such physical instruments as viscometers and X-ray cameras, is penetrating into the industrial field. The strength and elasticity of the fibre, its durability and suitability for dyes, can now all be made to specification. This is because the mechanism relating this performance to the molecular structure is beginning to be understood. The necessities of war had a decisive effect in speeding the development of the new chemical industry. The great synthetic rubber industry of the United States was built up in two years to serve the gigantic needs of modern war. This would have seemed inconceivable in peace-time, but the essential difficulties had always been financial rather than technical.

MOLECULES MADE TO ORDER

The era of polymers and plastics is only beginning, and they themselves are only the first examples of materials made to specification. What has happened in industrial chemistry is that by the application of science, particularly physics, it is becoming possible to equal and to surpass the natural products in performance and cheapness. The chemical industry, which as we have seen was brought into being largely by the textile industry, now seems set to replace it, at least as far as the production of fibres is concerned. This does not imply by any means that the factory will replace the farm, but it does mean that in future mines, farms, factories, and laboratories must be linked in one complex chemical production flow-sheet in which molecules are taken in the forms cheapest to produce and embodied in materials and articles capable of best satisfying human needs.

A SCIENTIFIC CHEMICAL INDUSTRY

The degree to which this has already been achieved is one sign that the chemical industry has become a truly scientific one, with an importance comparable only to that of the electrical industry. The difference between these two industries is that while the electrical industry was scientific in its very inception, arising entirely out of the eighteenth and nineteenth centuries' discoveries of electricity, the chemical industry has had to achieve a change-over from the most ancient traditional procedure to one based on a rational approach to the solution of definable problems. In both cases therefore the demand for scientists at all levels for future research, development, and production is greater than that of all the other traditional industries, not excepting the heavy and engineering industries. In fact, something like three-quarters of all scientific workers in industry are in the electrical or chemical industries.

THE FINE CHEMICAL INDUSTRY

Quantitatively, by far the greater amount of production in chemistry is in heavy chemicals and plastics, and is achieved increasingly by automatically controlled synthetic processes. Qualitatively even more important, at any rate for the future, is the fine chemical industry, which is tending more and more to become part of the new biology. The chemical techniques evolved in the latter part of the nineteenth century to deal with the commercially valuable dyes have now been very largely switched in the direction of the study of substances of biological importance, first in research and then very rapidly for full-scale use in medicine or agriculture. The scientific situation in biochemistry will be dealt with in Chapter 11. From the chemical point of view it is sufficient to say that the subject shows all the signs of being in its earliest and most rapidly developing phase.

SOCIAL NEED AND SCIENTIFIC PLANNING

The advances of biochemistry and chemotherapy have, however, shown that on this side science is going to be more immediately effective in human affairs than it has ever been in the past. The whole world can be changed more quickly now by some chemical discovery, such as that of paludrin for the treatment of malaria, or atricide for nagana, than by the addition of the energy of all the uranium in the world.

256. The new biology, with its elaborate chemical techniques, requires specialized conditions for much of its work. A sterile laboratory at the Laboratory of Embryology and Tetratology at Strasburg, which is basically concerned with the development of all living things, normal and abnormal.

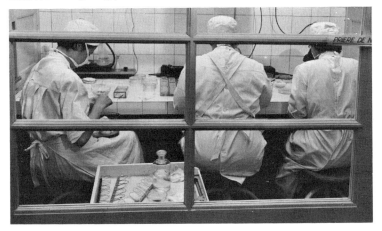

This fact makes the relative development of different branches of science a matter of urgent public concern. It can no longer be left to the personal inclinations and unaided efforts of individual scientists, often working in ignorance of the wider implications of their subjects, to develop this or that field at their pleasure, when the difference between developing one or the other may be a matter of life or death for hundreds of millions of people. This does not mean a need so much for directing scientists, but rather for a better system of scientific education adapted to a society consciously aimed at securing the maximum human welfare.

## 10.9 Natural Resources

### POWER, SOIL, AND MINERALS

Nowhere does the need for such a comprehensive view appear more clearly than in the use of the natural resources of our planet – rock and soil, water, air, and sunshine. These are the fields of the earth sciences which, though themselves engendered by man's experience in extracting natural riches (p. 640),[6.86] have remained until very recently largely descriptive and interpretative sciences. What has happened has been a haphazard and wasteful exploitation of mineral wealth, coupled with an even more dangerous interference with soil and vegetation. Until the present century this waste and destruction was limited and local. Now the increasing scale and power of machinery, and the even more rapid increase in the utilization of fuels and metals, threatens to destroy irreparably the slowly accumulated natural stocks of the whole planet.

So far, under capitalism, only ignorance has served to protect them. An earth which is private property, divided into spheres of interest of monopoly combines, whose interest may here dictate the wasteful and ruthless extraction of a natural resource to make quick profits out of low wages, and then refuse to develop it at all for fear of increasing supply and lowering prices and profits, cannot be known scientifically or used rationally. Actually, all over the so-called free world, knowledge of natural resources is limited to spotty prospecting carried out by companies, and to gentlemanly and extremely parsimonious official surveys carried out by governments deeply concerned not to interfere with private interests. Before the war they had discovered only a very small fraction of even the easily accessible minerals, as is shown clearly enough by the rich strikes that somewhat better supported surveys have made since the war.

The same lesson can be learned from the experience of the Soviet Union and of China. There the known natural resources, revealed by intensive surveys manned by rapidly expanded scientific teams, have increased far beyond the most optimistic of earlier predictions. The search for natural resources is being pressed in these countries, involving the training of geologists as a first priority, because on their discovery depends that of the most economic location for industry.[6.17] In opening up a large country millions of pounds of capital can be sunk or costly hauls of materials may be made necessary for lack of adequate knowledge of where the raw materials can be found in quantity. Thus the study of natural resources is closely linked to their utilization. It is now becoming apparent that the raw resources of Nature are not things merely there, to be ignored or picked up as such, but are calling out for integral human control. The mineral wealth, the water supply, the biological possibilities of the soil, the capacities of its inhabitants, are not something given and unalterable, but need to be transformed in a way that will secure not only the best use of each but also the best combination of all. In this, Nature has not only to be known and used but also to be transformed. The new mechanical, chemical, and biological possibilities offered by science are, as the experiences of the Soviet Union and of China are beginning to show, that rivers can be turned into chains of lakes or reversed, if need be; that plains can be afforested and deserts reclaimed (pp. 965 f.). Man can now work on the scale of Nature, and so multiply the resources previously available.[6,200] The transformation of Nature is necessarily as much a biological as a physical problem, so that fuller discussion of it is best deferred to the next chapter. Here it is enough to point out that with the increased scale of operation it is no longer sufficient to limit knowledge and activity to separate localities. Action needs to be on a world-wide scale. Even for complete knowledge of any one part of the globe it is necessary to use observations made all over it. For the full use of science in making natural resources available to all, international co-operation is more than ever necessary. The first or survey stage of such co-operation was started under the aegis of United Nations organizations, especially by UNESCO. Co-operative studies have been undertaken on the arid zone and the humid tropics. Co-ordination of observations and research into atmospheric phenomena, particularly in Antarctica, were undertaken for the International Geophysical Year in 1957.

## 10.10 War and Science

Unfortunately in this century, when international co-operation in science has been most needed and most useful, it has also been most hindered. Wars and revolutions, and the threat of still more to come, have been most effective in holding up the advance and diverting the uses of science.

Any attempt to deal with the growth of science and its relation to industry in the twentieth century must include explicitly the effect of war. Though, as has been shown (pp. 218, 320 f.), war has had an important influence on science in earlier centuries and has itself been modified by science, its effect has now become of an altogether different order. Many circumstances have combined in our time to make this so. The earlier applications of science to productive processes have contributed strongly to the economic and political imbalance which has given rise to the imperialism, crises, and wars of our century. Ten years have been spent in actual hostilities, in which the whole effort of industrial countries has been turned to perfecting, developing, and manufacturing new weapons, and at least twenty more in preparation for war, wherein the same activities are carried out at a slightly slower pace. The physical results of this are for everyone to see, at least in the old world, in the utter destruction of scores of cities that had taken centuries to build, in the less obvious but far more crippling and lasting human losses, and perhaps worst of all in the mentality that has been schooled to regard all these things as inevitable.

DESTRUCTIVE WEAPONS

The means by which this destruction was effected were largely scientific. Even before the atomic bomb, thousands of scientists had been employed and tens of millions of pounds had been spent by governments on perfecting aeroplanes, bombs, and radar navigation, to say nothing of the lethal improvements of older weapons. It is now abundantly clear that the use of physical science in this way has already done enough damage to hold up civilization for decades, and is capable, if pressed forward as it is being at an accelerated pace, of wiping out all

257. The application of science for destruction was seen, for instance, in the use of special fire bombs in the Second World War. Firemen at Eastcheap, London.

life over much of the earth. The threat of the hydrogen bomb has brought that home to the whole world.

POTENTIAL UTILITY OF MILITARY EQUIPMENT

The experience of war science points, however, equally strongly to a different and indeed hopeful conclusion. The very urgency of the needs of war has shown how much more rapidly physical science can be pressed forward and applied than anyone could be induced to believe in peace-time. Now the utilization of science, even in war itself, has been only to a minor extent for exclusively combatant purposes. Most of it consists in satisfying the same needs as exist in civil life, but doing so without delays and without regard to cost. The major technical developments in war are in the fields of communication, transport, and production. The walkie-talkie, the bulldozer, the D U K W, and the jeep are just as characteristic of the Second World War as the self-propelled gun, the super-fortress, and the atom bomb itself. The reconstruction of the world, and the extension of civilization to previously barren regions, can be achieved far more rapidly with these simple and useful devices than was ever thought possible before. DDT and penicillin, though not themselves originated by war research, have been developed and used to an extent that would have been impossible had the war not taken place.

Even in the development of weapons themselves, the method of application of science may be essentially the same in peace and war, except that in capitalist countries it is pushed forward in full vigour only under the impetus of the fears of loss and hopes of profit that war alone can bring. It is also only in war that a high degree of planning and consideration of over-all effects can be reached. All these aspects are apparent in the cardinal scientific developments of the Second World War, notably in that of radar.

OPERATIONAL RESEARCH

It was not only in the field of production of weapons that the experience of the war was to add to the range of action of the physical sciences. For the first time, in war, the work of the scientist took him from a consideration of the weapons to that of their uses on the field of battle. From the result of these studies it was almost inevitable to go on to the scientific treatment in observation and experiment of actual military operations, on land and sea and in the air. *Operational research* has been defined as 'the use of scientific method, particularly that of measurement, to arrive at decisions on which executive action can be based'.[6.103]

It was used extensively and often decisively, for example in the anti-submarine campaign,[6.48] by the British and later by the American forces; it was not used by the Germans, and this omission contributed to their defeat, both in their failure to find counters to enemy weapons and in their expending disproportionate efforts on weapons which operational research would have shown to be useless.

The Soviet armies did not, as far as we know, use any separate corps of operational research. This would not have been necessary for them, for owing to the radically different class composition, training, and traditions of the Red Army, science was from the start an implicit part of their operational training and action. The achievements of that army both in the production of superior weapons, old and new – tanks, guns, and rockets – and their use in the field show the degree to which science can be used in warfare, flexibly and with imagination. It is usually forgotten that the use of paratroops, now thought of as an invaluable adjunct of attacking forces, was a Soviet innovation and was thoroughly ridiculed by foreign military experts when it was first tried out.

It was only in its inception that operational research was exclusively confined to the physical sciences. Because it started with such gadgets as radar and bomb-sights, operational research men tended to be physicists. Essentially, however, the method was one of human organization, and will be treated as such in Chapter 14. Its importance in this context is that it was the first way in which physical science, engineering, and full-scale practice were joined together in one conscious common discipline which has far wider implications than those of war, particularly for industrial production.

## LESSONS OF THE ATOM BOMB

The supreme example of the production, exclusively for war purposes, of a scientific discovery within the hitherto incredibly short space of three years was the atom bomb. As a scientific and industrial enterprise this development represents the most concentrated and, in absolute figures, the greatest scientific technical effort in the whole of human history. In fact the sum spent on the atomic project, some £500 million, is much more than had been spent on the whole of scientific research and development since the beginning of time, but it is already completely overshadowed by the cost of rocket developments.

Under any rational system of the utilization of science, on the other hand, atomic fission would have been the centre of the most intense development, leading to its use for the production of power and for the other uses to which the products of the pile could be put (pp. 759 ff.).

258. The most significant use of basic scientific research in the military sphere has been in the development of nuclear weapons. The power of a nuclear device may be gauged from this underwater explosion made in the lagoon of Bikini Island in the Pacific, July 1946.

Actually, as we all know, it was developed for a different purpose, that of producing a bomb and wantonly killing at Hiroshima 60,000 and at Nagasaki 39,000 people. This act had no military justification. Even in the official *Report on the Pacific War* we find the statement:

... Based on a detailed investigation of all the facts, and supported by the testimony of the surviving Japanese leaders involved, it is the Survey's opinion that certainly prior to the 31st December, 1945, Japan would have surrendered even if the atomic bombs had not been dropped, even if Russia had not entered the war, and even if no invasion had been planned or contemplated.[6.24; 6.36]

The very existence of the atom bomb, the threat of its use by the United States against its former allies, the tragic farce of spies and secrets that were no secrets, have done more than any other products of science to embitter international relations and to spread terror and despair through the world. In the United States the knowledge that the Soviet Union had also developed an atom bomb intensified the feeling of suspicion of which the Rosenbergs were the victims. The counter-move was not to agree to prohibit all atomic weapons but a rush to develop the far more terrible hydrogen bomb (pp. 839 f.). It was opposition to this policy that was the real cause of Oppenheimer's (1904-67) fall from grace. (See p. 1277.) From the outset atom-bomb developments had important effects on science, economics, and politics.

In the United States particularly, the influence of the Atomic Energy Commission has been to divert research in an altogether unbalanced way in the direction of nuclear studies.

The whole subsequent history, the involved and till now inconclusive discussion on the abolition of the bomb and on the control of atomic energy, brings out as never before the key role of physical science in international politics.[6.24; 6.36] To this aspect we shall return later. Here it is sufficient to emphasize the new kind of large-scale industrial enterprise that has grown around atom-bomb production, implying a closer partnership than ever by monopoly electrical and chemical combines with the military and the Government, by which, without any risk to themselves, the firms can draw ever vaster sums from the Treasury. The proposals to extend the same system to Britain, embodied in the Atomic Energy Act, and now the various moves for Euratom and other means of financing atomic energy within the orbit of capitalism show the common tendency to bring the new forces, revealed by science, into the service of profit and war.

The history of the atom bomb also carries the lesson that even in the orbit of capitalism, under the threat of war, such a great enterprise co-ordinating different sciences and techniques can be put through on a planned basis. This furnishes an undeniable proof of what science could do, if it were strategically applied to the satisfaction of human needs rather than to purposes of destruction.

GUIDED MISSILES

The atom bomb was the most destructive use of science in the service of war, and it also made use of the most radically new developments of science, but it was not the only development of crucial importance. Of comparable significance were the applications in radiation physics and information theory, exemplified in tele-communications, radar, servo-controlled gunnery, proximity fuses, guided and homing missiles that came into service towards the end of the war, and have been intensively developed ever since. The principles underlying these developments have already been discussed (pp. 782 f.). Here it is only necessary to point out how much radio and electronic research was speeded up in the war effort, and how military requirements for light, compact, and above all expendable equipment transformed the manufacture of components and led to miniaturization, culminating after the war in the minute transistor taking the place of the bulky valve. Miniaturization itself, though a solution to the problem founded by war-time conditions, is proving one of these keys that unlocks far more

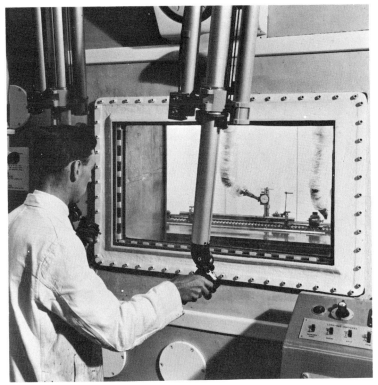

259. The danger of radioactive weapons lies not only in their direct effect and the danger of fall-out (see next illustration) but also in their ability to make surroundings themselves radioactive. In nuclear laboratories, equipment which is used for radioactive work has to be remotely handled to prevent danger to the operator. Photograph taken in the Berkeley Nuclear Laboratories, California.

locks than the one it was designed for. Small size goes with increasing rapidity and action, and makes not only for the minor horror of the transistor radio set but also for all kinds of measuring methods which were not achievable at all before, as well as for computers and for microtransmitters in physiology and medicine (p. 782).

## SCIENTIFIC AND INHUMAN WARFARE

The ultimate purpose of the introduction of electronic control and direction into weapons is to remove the human element in warfare still

260. The use and test of nuclear weapons, as well as the generation of electricity in nuclear power stations, necessitates the monitoring of fall-out of radioactive particles from the air. Routine Health Physics Surveys are made, for instance, in the neighbourhood of the Dounreay Experimental Reactor Establishment.

farther from the actual combat zone, or more crudely to ensure the safety of the weapon wielder by keeping him well away from the results of his work and from immediate retaliation. The use of such weapons does not, in fact, make war any more humane. Far more physical pain and suffering were caused in a shorter time to more people by the Americans and their allies in Korea by the use of high explosive and napalm[6.121] than in any comparable war in the past. These weapons do, however, add enormously to the cost of warfare, and limit the ability of waging war to highly industrialized States making full use of science or to their less highly industrialized and more expendable clients.

Further, the great gap between action and result fosters the irresponsibility of modern warfare, which in its thoughtlessness rivals in reality the deliberate cruelty of earlier ages. Professor Nef in his *War and Human Progress* presents a well-documented history of the progressive degradation of decent behaviour in warfare, keeping step with the improvement of lethal weapons. Push-button war permits well-meaning and apparently civilized men to perpetuate with a clear conscience the most ghastly massacres, the effects of which they never see.

Even more dangerous is the belief in the efficacy of push-button war, for it is in this belief that aggressive or so-called preventive war is lightly entered into. The Trenchard–Douhet theory of winning wars by destruction of enemy war potential by strategic mass bombing was responsible for most of the devastation of the Second World War without producing any decisive strategic advantage.[6.127] German war production actually increased under intensive British and American bombing.[6.24] Despite this the idea is more prevalent than ever. Armed with the atom or hydrogen bomb and guided weapons, some dangerous lunatics in high places reckon on winning World War III in a few hours or days. Misplaced confidence in that belief may well touch it off (p. 842).

THE HYDROGEN BOMB

These ideas, drawn from pre-atomic days, have produced their own Nemesis in the construction of the hydrogen bomb. Once the bomb race was on, it seemed that the side that first got to the hydrogen bomb with a destructive force of a thousand times or more that of the 'conventional' atom bomb would acquire decisive advantage and, as the Americans openly boasted, command an unshakeable 'position of force' from which to negotiate. Actually in this case, unlike that of the first fission bomb, the United States seem to have had no initial lead, in fact they may have been slightly behind the Soviet Union. This aspect of lack of decisive advantage, reinforced by the very extent of the damage caused by the hydrogen bomb, led to the beginning of the period of nuclear stalemate which in various forms has gone on ever since.

At first the hydrogen bomb was thought of as simply a much more powerful atom bomb, the power of each explosion being measured by the tens of millions of tons (megatons) of TNT needed to match it. As such it was evident that one would be sufficient to wipe out a large capital city such as New York or Moscow or an industrial district like the Ruhr. As it would be a waste to use it on smaller targets, its use implied the theory of the decisive blow and the thirty-six-hour war.

However, it was another effect which made even this degree of destruction seem unimportant and introduced a new and apparently ultimate horror into scientific warfare. This was the *radioactive fall-out* which came to light by a tragic accident following US tests of hydrogen bombs in the Pacific on 1 March 1954. A shift in the wind exposed some Japanese fishermen seventy-five miles away to radioactive ashes causing severe radiation sickness, and one death six months later. The nature of the radioactive material as analysed by Japanese physicists showed, surprisingly enough, a preponderance of uranium fission products. It was at once apparent that the bomb owed its power to a triple and not a double nuclear reaction. The neutrons from hydrogen fusion explosion were being used to break a uranium case round the bomb. Now that it was useless to keep the secret the US authorities released data about the fall-out which showed that its lethal potentialities were greater even than the direct effects of the explosion. From such a bomb – released within a few thousand feet of the earth – a completely lethal concentration of fall-out might be expected over some thousand square miles, and up- and down-wind of the explosion something varying between four times to half the lethal dose for an exposed person over a further six thousand square miles, distributed down-wind in a long streak some two hundred miles long. The qualification of exposure is of little value, for much of the radiation is of long life, and those under cover at the time of the fall-out could hardly escape uncontaminated from the area. Nor is even this the whole story. Some of the radioactive material is carried high into the upper atmosphere, distributed all over the world, and takes years to fall down. No living thing can escape. Sheep in Wales already show in their bones strontium 90, derived from the Pacific explosions. These weaker radiations also have genetic effects and may produce monsters for generations (pp. 953 ff.).

It has taken some considerable time for these facts to be appreciated by the minds of ordinary people and even longer by those of politicians and military men. They turn any prospect of war into one of general slaughter, inconceivable in its extent and suffering and virtually un-limited in its after-effects. It is unlikely, despite certain authoritative predictions, that everyone will be mercifully killed, but few would escape radiation sickness and none genetic effects. Half a dozen bombs would wipe out the cities and industry of Britain. An American army expert even claimed that the all-out US bombing effort would kill some 700 million people, one quarter of the world's population, but that the deaths would by no means be limited to enemies.[6.35]

Indeed, those US military theorists are prepared to accept up to sixty million, or sixty mega-deaths, in the United States itself as a reasonable number which would still justify waging a nuclear war against the Soviet Union to defend vital US interests.

THE MISSILE AGE, COUNTER-FORCE STRATEGY, OVERKILL

By 1957, with the appearance of the first sputnik, it was evident that there were possibilities for intercontinental ballistic missiles and this changed the whole aspect of modern warfare. The concept of the long-range rocket, itself in the first place a purely conventional weapon (p. 813), has completely altered the character of modern war. This, combined with the efficacy of control systems and interception systems, has put an end to the long supremacy of the aeroplane as a weapon carrier. The intercontinental rocket and the super megaton hydrogen bomb are, indeed, inseparably linked. Long-range rockets would, of

261. The development of space research is one of the aspects of the missile age, intercontinental ballistic missiles are another. The Polaris A-3 missile.

course, be entirely useless in the pre-atomic or even in the fission bomb stage because so many would have been necessary to do a serious amount of damage. But the advent of the hydrogen bomb changed the whole picture and it is now evident that even a relatively small number of hydrogen bombs, in fact considerably less than those already in existence, would be sufficient to wipe out not just cities but whole industrial districts, not to mention the more extensive damage due to fall-out.

At this stage this means a condition of effective nuclear stalemate between the great powers. At an intermediate stage it seemed that this difficulty could be met by the spreading of nuclear rocket bases over a large portion of the world, as indeed was done by the United States with its over two hundred overseas bases. These were then far too vulnerable and the attempt to get out of the difficulty has been met by the idea of protecting the bases either by hardening them, that is, putting them into heavy concrete shelters, or by making them undetectable by mounting them in submarines, the Polaris nuclear submarine system. The effect of this on strategy has been revolutionary. Inspired by the new scientific or 'game theory' strategists, such as Professor Teller, the 'father of the hydrogen bomb', and supported by the military research corporations of which the most famous is the Rand Corporation and its most voluble protagonist Mr Kahn,[6.90] a new strategy has been evolved called the 'counter-force strategy' in which the first strike is aimed at crippling the striking capacity of the other side with the object of winning the war without losses. This counter-force strategy realizes that all the hidden weapons will be difficult to knock out and therefore it will require a large number of nuclear bombs to do so. This requires a much larger production of bombs, what is called 'over-kill' capacity in spite of the fact that there is already enough nuclear material stored and capable of being delivered to kill everybody in the world ten times over. The fact that the ordinary population near the hardened sites would also be killed and further destroyed by fall-out is called a 'bonus' kill.

In 1962 this strategy was officially adopted by the United States Secretary for War, Mr Macnamara, and it has certain advantages in peace as it leads to far greater orders for missiles which can, to a large extent, offset the losses due to cancelling military aeroplane orders. It is considered a humane policy because it does not aim at cities as the so-called deterrent policy did which preceded it.

All this depends on whether the Soviet Union intends to play the game according to the rules laid down by the Pentagon's 'games theory' men or whether, as has been often stated, they consider the only solution to this problem is disarmament and dismantling of these weapons, beginning with the Polaris and all nuclear weapons and passing on from that to total disarmament.

The whole of this scientific nightmare seems hardly worth recounting if it were not the real state of affairs under which we now live and if it were not taking up the time, energy, and imagination of hundreds of thousands of scientists, the largest proportion of those in the most important fields, those in electronics and chemistry.

These weapons are probably not the latest word. Already various types of anti-missile missiles have been developed; they give advantage to the defence, to be countered later by anti-anti-missile missiles.

Outside the small group of the old nuclear powers to which there has been one addition in recent years – that of France, with its somewhat outdated Force de Frappe – there is widespread repugnance to nuclear war and nuclear weapons, but it has been so far quite ineffective. Certain lip service has been paid to the ideal of universal disarmament, and indeed there exists a reasonably good and mutually acceptable plan for this, but until the forces that maintain the armaments industry in the United States and have immense powers in the Senate, the military-industrial political complex which even President Eisenhower denounced in 1960, lose their power, there is little hope of escape.

## EFFECTS OF WAR ON SCIENCE AND SCIENTISTS

The great change in the status and place of science brought about by the Second World War has already been alluded to. It is primarily physical science that has been affected, because it was the most advanced and the most closely linked with war and industry. It was especially in physics that the war led to the greatest interruption in Britain and America. Most academic laboratories were closed or turned over to war uses, and the most brilliant men occupied themselves with problems which had no relation to their previous work. The supreme importance which the war gave to physical science, mainly in relation to atomic energy and electronics, has lasted into the post-war period. This has meant, particularly in the United States, a great expansion of physical research and its equipping with vast and expensive apparatus such as experimental piles, synchrotrons, and electronic calculators.

## THE DOMINANCE OF MILITARY SCIENCE

Now these and the general scale of the work are entirely beyond the scope of even the wealthiest universities, or even of industrial firms, and they can therefore only be found either in special government laboratories or in government-subsidized university and industrial laboratories. Actually, both methods have been followed, with the result that government laboratories have come to rival universities for postgraduate work, and university physics departments have become annexes to government contract schemes inside them. In itself this would do little harm and might do good, by bringing universities more into contact with contemporary engineering practice, were it not for the fact that all this research is supported ultimately for its future military value. The effects of war have been to consolidate still further the predominance of government and monopoly control of physical science. The degree of control varies widely in different countries. It is most thorough in the United States.[6.25; 6.29; 6.31; 6.76; 6.90; 6.93; 6.97; 6.109]

From the outset, all works of a nuclear character and all those dealing with control mechanisms and rocket developments have been subject to stringent security restrictions. This goes much further than mere control over research results that might be of military value. It affects the whole life and thought of all scientists whether in universities or institutes. It implies loyalty oaths under conditions where refusal to swear means dismissal and smearing as a security risk. Smearing puts the scientist at the mercy of any informer who may allege his connexions with any of a whole host of subversive organizations or by mere guilt of association. Witness the tragic case of Dr Robert Oppenheimer, who was belatedly rehabilitated. The effect of this atmosphere on new generations of scientists is to discourage any independence at all, indeed any thought outside their narrow scientific speciality, particularly any thought of social or moral responsibility for what they have done or are doing.

In Britain the share of government rather than of universities in military science is much greater, and strictly secret research is almost, though not entirely, confined to government laboratories. The connexion with big business is also far more indirect, and the finance of the universities, though it comes mainly from the Treasury, is administered by university men. In this way the grossest evils of the militarization of science are avoided, but at a cost of cutting down the development of research. Further, the influence of thought control is exercised in a way much more subtle than in America, but one more difficult to react against. Teachers are rarely dismissed for their political opinions, they are merely not appointed to important posts if their views are regarded as not sound.

In France the situation is again different, owing to the complications of the Occupation and the large and influential Communist party. Many prominent scientists are excluded from leading positions for political reasons. In 1950 Joliot Curie, one of the discoverers of nuclear fission and a noted Resistance leader, was removed from his post of High Commissioner for Atomic Energy because he said publicly that atomic energy should never be used for war.

Despite this official pressure, scientists throughout the world have persistently protested about the dangers of atomic warfare and the misuse of science for war. Associations of atomic scientists in the United States, Britain, and France have kept up a well-informed pressure on this subject, while more general bodies, such as the World Federation of Scientific Workers and Science for Peace, and other scientific organizations in India, China, and the Soviet Union have also added their voice. In Japan practically every scientific organization has been

262. The development of nuclear physics involves such expense that some measure of international co-operation is needed. The great proton synchrotron near Meyrin, Geneva, is operated by C E R N (see plate 225). It has a circular synchrotron evacuated chamber 656 feet in diameter yet built to the most exacting instrument-making specification. It is beyond the ability of any European country to finance such a project alone. The photograph shows part of the magnet system which provides the guiding field for high energy nuclear particles.

campaigning against atom bombs and bomb tests. The two most well-reported manifestoes were that initiated by Bertrand Russell and Einstein demanding an end to war, and a more limited appeal from the majority of scientific Nobel Prize winners.[6.122]

### THE PUGWASH MOVEMENT

The first of these manifestoes, however, was to have a permanent and growing influence. It was to lead to the convening, first at Pugwash in Nova Scotia, thanks to the public spirit of Mr Cyrus Eaton, a railway magnate, of groups of scientists from the major countries involved, including the United States, Britain, and the Soviet Union, to discuss the actual situation of scientists in the face of the threat of nuclear war and their responsibility for it.[6.37; 6.120]

Over the years since 1957 there have been more than ten such meetings and considerable work and publication in between. Limited at the outset to scientists who already had a social conscience and who had demonstrated their willingness to agitate for the use of science for peaceful purposes, they became more and more semi-official exchanges, including just those scientists who were occupied in military science on both sides of the Cold War. This had an undoubtedly positive value because the conferences themselves were able to act as reconnaissance patrols for agreement on disarmament and, indeed, particularly in their ninth and tenth meetings in England, they worked out a reasonable and agreed scheme for disarmament by stages, a compromise between those put forward by the United States and by the Soviet Union. It sponsored, for instance, the development of the automatic 'black box' seismic station later adopted officially by both the US and USSR government in their negotiations for stopping all nuclear tests.

They also considered, though possibly not to the same extent, the possible positive uses of the scientific effort at present wasted on war for constructive purposes, particularly for help to the underdeveloped countries. Though all this has had its effect on public opinion, it is evident that far more will have to be done before all the governments concerned are seriously prepared to agree to the banning of nuclear warfare.

### THE COST OF MILITARY RESEARCH

The bitter fact remains that, to a degree unimagined before, physical research in capitalist countries is becoming dominated by military demands. In the applications of research the military aspect is even more prominent. The amounts devoted to specifically military research and development in the United States and Britain are now many times greater than those spent before the War, as the following table shows:

EXPENDITURE ON RESEARCH AND DEVELOPMENT (£ MILLIONS)

| | Industry | | | Government | | | | | |
|---|---|---|---|---|---|---|---|---|---|
| | | | | Civil | | | Military | | |
| | 1937 | 1955 | 1962 | 1937 | 1955 | 1962 | 1937 | 1955 | 1962 |
| USA | 61 | 920 | 1,800 | 20 | 140 | 960 | 5 | 710 | 2,800 |
| Britain | 3 | 65 | 213 | 3 | 36 | 139 | 1·5 | 214 | 246 |

The post-war expenditures completely dwarf those for civil research and development over the same period. How the money is spent it is more difficult to say in view of the cloak of secrecy that surrounds it. It is probable that the lion's share goes to armaments firms, including chemical and engineering, for new substances and component mechanisms. An altogether disproportionate amount is likely to be spent on the development of mass-destruction weapons and of means for projecting and controlling them over large distances. It may well be that the money is largely wasted, as is the habit in military establishments, with their predilection for full-scale trials from which very little can be learned and with their immunity on the grounds of 'security' from scientific or economic checks.[6.11] Though conscience has kept a great number of scientists out of military science, among them Kapitza in the Soviet Union and Urey in the United States, and security checks have kept out others, enough are left in to mark a large potential loss to science.

The hundreds of millions of pounds or thousands of millions of dollars which have been made available to military research indicate clearly enough what could be enjoyed by civil science under a saner system. It would in fact, if wisely distributed between education, research, and development, completely transform the situation for science and make possible an enormous leap forward in the speed and value of its application to the satisfaction of human needs. Such a use of science in a capitalist society, however, is hardly to be expected. The reasons will be discussed in Chapter 14; it is sufficient to say here that while government-sponsored research for peace interferes with the exploitation of the consumer by private or monopoly industry, government-sponsored research for war brings them development contracts and ensures profits without risks. [6.19; 6.109]

For the same reason the major lesson of the war for science – the value of strategic planning – cannot be transferred to peace-time conditions (pp. 833 f.).[6.11] What the war showed was that it was possible to sort out problems, even fundamental problems, over the whole field of the war effort and to give them some order of priority. This was done in relation both to the importance of solving them and to the chances of getting them solved in a reasonable time, taking account all the while of the qualifications, personalities, and interests of the scientists available from each discipline. Some such strategy and planning in peace-time science is needed now more than ever before, but whether it can be achieved is not a problem of physical science but of society itself, the discussion of which must be left until the fields of biological and social science have been surveyed.

## 10.11 The Future of the Physical Sciences

Before passing on to these fields it is worth while to examine for a moment what the future may hold for the physical sciences and for the productive industries with which they are so closely linked, and to consider the contribution that the physical sciences may make to the thought and culture of the coming years. Social and economic factors may, and indeed in the last resort must, control the over-all speed with which science and industry advance. They may also, though to a lesser extent, determine the direction or distribution among the sciences of the total scientific effort. Nevertheless it remains true that science and industry can only proceed in the short run on the basis of existing equipment and ideas. Revolutionary discoveries and theories may quite unpredictably alter this picture, but not all over or all at once, as is easily seen by considering the time it took for the quantum theory to make itself felt. Even atomic fission, for all the billions of dollars spent on it, has made as yet little overall difference to the course of physics.

Nevertheless it would be idle to attempt to forecast separately the future of fundamental and applied science. The combination of research and development in physical sciences is bound to become ever closer, with the role of fundamental science ever increasing. Engineering is in process of being rapidly transformed by the sciences that brought it into being three centuries ago.

The lead in technological change will henceforth remain with science. The era of rule-of-thumb change is over. Further, the rate of transfer between discovery and production is itself rapidly increasing. In every field of physical science each new scientific development is likely to be embodied within a few months in practice, and new practical experience will react back, though usually not quite as quickly, to provide fundamental science with new instruments and new problems.

Another aspect of the same general tendency to coherence is the increasing interrelatedness of the different scientific disciplines, reaching in fact far beyond the physical sciences into the region of the biological and social sciences. With it goes an ever greater need to understand the whole pattern of scientific and technical effort, so that it may be able to organize itself in a strategic advance and not be dissipated in unconnected raids into the unknown.

To turn from these general aspects to particular developments, it is possible, without making any attempt at detailed prediction, to point out some marked trends along which notable advances and applications may reasonably be expected. This is not to say that the most exciting discoveries will be along these lines, the very admission of their being exciting presumes their unpredictability. With a given effort of science we may be certain that they *will* occur, but *where* or *when* we have no means of knowing. Certain fields rich in recent discoveries may for the time being be worked out, others where scientists have marked time for decades may be on the verge of revolutionary changes. Nevertheless, experience of advances in the recent past may not go for nothing in predicting the near future.

THE FUTURE OF NUCLEAR PHYSICS

In view of its intrinsic importance, the nature of fundamental particles and their interaction in low- and high-speed encounters and in more or less stable nuclei takes pride of place in physics. If we add to this its immediate military importance and its possible later exploitation for industrial purposes and the billions of dollars spent in such research, it is here that we can expect the greatest advances. The very confused and self-contradictory state of particle and nuclear physics is a sign that a new and comprehensive theory may emerge. Indeed it is already long overdue. The new arsenal of experimental devices, on the one side accelerators and piles for producing particles, and on the other scintillators and counters for detecting them, are bound to produce new data and phenomena which may stimulate theoretical physics and even provide clues for a long overdue advance (pp. 862 f.). In several ways it will help to advance the other sciences such as chemistry and biology through the use of tracers, and in a more social sense by forcing a general increase in scientific education and research.

The moves towards international co-operation in nuclear physics were at first one-sided, resulting in the provisional setting up in 1952 of the Centre Européen pour la Recherche Nucléaire (CERN), formally incorporated in 1954 with the co-operation of twelve European governments, but with an essentially scientific direction. The main experimental establishment is in Geneva, where a number of particle accelerators are being built; the largest, of 29 BeV, started operation in 1960. Proposals to make the US centre of nuclear research at Brookhaven a

263. High velocities must be given to atomic particles for artificial disintegration of the nucleus if studies are to be made of its nature. This is achieved by, for instance, accelerating protons using a very high voltage to speed them through an evacuated tube and inject them into an evacuated annular ring – the synchrotron. Protons have been made to reach a velocity 99·94 per cent of the velocity of light. The long evacuated tube – the linear accelerator – at Stanford University in the United States is 3 kilometres in length.

common centre for governments in the Americas and Asia have also been made. In 1956 another joint nuclear-research centre, representing eleven governments ranging from Poland to Korea, was set up at Dubna. It also has large accelerators available to it, including the existing 10-BeV Phasosynchrotron.

Both centres are pledged to study only peaceful effects of nuclear energy. Though at the moment no State adheres to both centres of research, there is already some measure of informal co-operation between them. It is to be hoped that this will get closer and more official as time goes on. Serious research in nuclear physics is now beyond the resources of most nations, and they can join in it only by such co-operation.[6.112]

ATOMIC POWER

As already discussed (pp. 759 ff.) atomic power for civil uses has arrived not only for generating electricity but for transport. The Americans have atomic submarines. The Russians have an atomic ice-breaker that can melt ice as well as crash through it; there is talk of atomic locomotives and aeroplanes. Atomic rockets may provide the key to long-range space travel. Within half a century, if war can be outlawed, cheap power from nuclear fission and possibly even cheaper power from nuclear fusion will be available in quantity. This means a potentially unlimited supply of materials and food. Power can be used to extract every kind of metal from its ores, however poor. Steel and aluminium will be as abundant as we want. Plants can be grown over the whole desert belt with water pumped up or distilled from the ocean, or flourish in hot houses through the long Arctic summer. With energy as free as air, there will be other limiting factors to contend with, but none are likely to prove as difficult to master as those social factors which are a legacy of the age of scarcity, inequality, and exploitation (pp. 1212 f., 1308 ff.).

UNDERSTANDING, EXPLORING, AND UTILIZING THE UNIVERSE

One very clear trend in modern physics, much enhanced by the study of the atomic nucleus and cosmic rays, is a renewed interest in the outer universe, planets, stars, and galaxies. This is the scientific aspect, so curiously and unfortunately mixed with the military, of man's entry into the space age. It is to be hoped that when the conflicts arising out of the Cold War have been damped, it will be possible to carry out a co-ordinated exploration of space both by manned and unmanned vessels on a world basis. In the meanwhile we must look at the space age as by no means an unmixed blessing for science. The sums of money used up on it are out of all proportion to the results obtained, compared with the spending of far less money in other sciences.

Nevertheless, in more ways than one, the study of outer space has had and can have far more effects on the development of science. It has, for instance, emphasized the importance of plasma, the high-temperature mixture of atoms and electrons moving in electrical and magnetic fields, such as are already being explored in the production of thermonuclear energy (pp. 759 ff.). It is apparent, for instance, that magnetic fields, however feeble, can, if they extend for distances into galactic space, profoundly alter the motion of matter when it is in its plasma state and over long ranges be more effective than gravitation itself in collecting material in the form of wisps from which stars may later condense.

One aspect of this is that the laboratory itself may extend into space. The first steps of this have already been taken with the setting up of telescopes and television cameras on artificial satellites. Further examination of the outer universe by the new methods places even more stress on their understanding in terms of nuclear and fundamental particle physics. Indeed, the understanding of nuclear processes demands not only an attention to the structure of the outer universe – relative abundance of elements, cycles of energy changes in sun and stars – but also to its history. The entry of the historic element into physical science completes the connexion between it and biological and social science.

The optical, radio, and cosmic-ray telescopes now appear as the lookouts for a physical exploration and colonization of the universe by man. The idea of space travel and colonization has already captured the imagination of the young. Once it is fairly started on an international basis it will be such an absorbing adventure that it may well provide, to adapt William James,[6.88a] the technological equivalent of war.

SOLID-STATE PHYSICS

The importance of this new branch of physics is rapidly growing because of its services, on the one hand, to engineering and nuclear physics in producing metals and materials of new properties (p. 789) and, on the other, to electronics, in providing crystal oscillators, ferromagnetic powders, and transistors (pp. 789 f.), as well as luminescent and fluorescent materials. Rapid developments in the theory of crystalline bodies, the phenomena of their growth, and their dislocations and imperfections are at last bringing exact science into the field of industrial practice, and we may expect great new advances leading to substances with exceptionally valuable properties. The study of piezo-

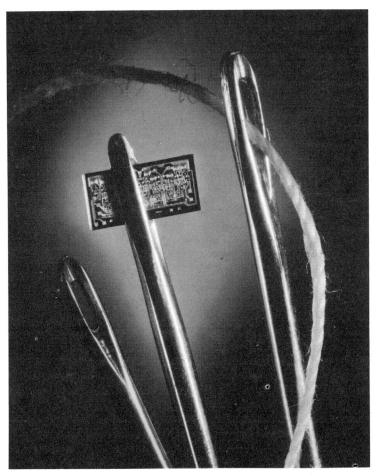

264. The advent of solid-state physics, the invention of the transistor, and the demands of military and space technology, have led to the development of micro-miniature circuits. The illustration shows, through the eye of a No. 5 sewing needle, a complete electronic counting circuit, made by Mullard and used by the thousand in modern computers. Known as an integrated circuit, it is shown here before being encapsulated in its mounting. The circuit is fabricated on a silicon chip measuring only 1·5 × 3 millimetres and yet contains over 120 components. If ordinary components were used to make the circuit, it would require an area of about 9 square inches – more than a thousand times that of the silicon chip. The degree of miniaturization can be appreciated when it is realized that the 'rope' is 40 gauge sewing cotton.

electric crystals used for crystal oscillators led to the discovery of substances of very large dielectric constants called ferroelectrics by
analogy to the ferromagnetic substances like iron. The new phenomenon
of anti-ferromagnetism, where the elementary magnets are arranged
pointing opposite ways rather than in the same way, has also been
discovered. The properties of semi-conductors have been shown to be
extremely sensitive to impurities – less than one part in a million is
detectable – and, by a purely physical method of zone melting, it has
been possible to achieve a purity far greater than any chemist has ever
demanded.

On the mechanical side great interest has centred round the production and behaviour of dislocations which are responsible for the plastic
yield of materials. It seems impossible to avoid them in bulk, but it has
been found recently that the hair-like crystals, in which many metals
and other crystals can be formed by suitable treatment, contain only
one long spiral dislocation and are extremely resistant to deformation.
This may prove a way to obtain tensile materials of an altogether new
order of strength.

THE THIRD INDUSTRIAL REVOLUTION: AUTOMATION

Power is blind; we use energy wastefully as a substitute, but an expensive one, for taking thought. Though the productivity per man hour has
increased in the last decade that per kilowatt hour has actually decreased. Merely to multiply man's material base without improving
efficiency and skill is to risk ruining our planet irreparably.

That efficiency and skill can now, however, be provided by further
developments of electronics, some of which have been already discussed
(pp. 782 ff.). The transformations now going on in industry, particularly
in mass-production industries, are not merely an extension of mechanization. We are justified in referring to *a new industrial revolution* because
of the introduction of the elements of control, judgement, and precision
that electronic devices can provide, as well as the greatly increased
speed with which industrial operations can be carried out. Automatized
production lines and even fully automatic factories are growing in
number and range (p. 817), but the full logic of the use of these devices
in all branches of industry still needs to be worked out. It is coming
quickly, for the leading ideas have been well grasped. What is still
holding it up, especially in capitalist countries, are economic factors
of vested interests and the shortage of scientists and technologists
(pp. 1153, 1189 f.).

Automation, though barely starting in capitalist countries, has already brought unemployment, and threatens more. To make full use of it implies a rational though flexible complete production system. The general flow-sheet of all industry and agriculture, transport and services must be maintained and continuously improved. This implies the use of calculating machines capable of handling the resulting complexity, which are rapidly being developed. The whole of economic life – wages, purchases, taxes, and pensions – could all be handled automatically, without the millions of desk slaves who now have to waste their lives dealing with them.

### THE FUTURE OF COMPUTERS

It is evident that we are at the beginning of the age of computers which are likely to have a bigger effect on the development of human society than any of the other products of the scientific revolution. The change has come about so quickly that few are able to gauge the range of possibilities it offers or to plan in advance how to deal with it. Indeed, computers themselves would be required to attack such problems.

The effect of new computational devices on mathematics, physics, and other sciences will, in the long run, be much greater. Not only will it make possible calculations hitherto far beyond human capacity, but it is certain to alter radically the whole of our thinking about quantitative methods in calculation in the same way as, and to a far greater extent than, the adoption of Arabic numerals did in the late Middle Ages. The new machines are no substitute for mathematical thought, rather should they stimulate it to new efforts (pp. 784 f.). Another aspect of electronics with almost unimaginable possibilities of extension in the future is the ability to translate and code any type of sensory data, and to change from one presentation to another, as radar and television screens do today. Already, electronic reading and speaking and translating machines exist, and there are possibilities of speeding up direct communications between minds, based on the physiology of the nervous system (pp. 942 f.). In America, Russia, and Britain prototypes of translating machines are already at work, and none too soon for, with the multiplicity of new languages in which scientific and technical information is produced at an ever-increasing rate, human translators can hardly be expected to keep up. Economy would seem to indicate that three or four such stations could serve the world with one master copy in a purely symbolic or numerical code – Bishop Wilkins' universal character. There would be no pretence to literary excellence, but there must be some check against distortion of essential meaning. Ultimately, miniaturized, portable, speaking and translating machines combined should make conversation between persons of different languages possible.

All such devices are, it seems to me, just in those stages of an even greater transformation: one which will release man altogether from the limitations placed on him by the major difficulties of organized thought and communication. Man arose as a social animal through the evolution of language. He was able at a far later date to fix some kind of collective memory through lasting writing. Merely to use electronic tools to improve these is to underrate their capacity. They should be conceived of as a substitution for thought and not merely for communication. In the first place thought itself might be in some way exteriorized and coded, and thus the limits of personal memory be transcended. The terrible waste which at present occurs in the perfection

265. Computers can design computers although such designs are as yet in the experimental stage. Photograph by Bell Telephone Laboratories.

of the human being by long education and then the loss of the resulting complex through merely physiological death could be offset to a far larger extent than it is today by the mere writing of books and memoirs.

But beyond that there is another possibility for computers that we are only just beginning to feel. Already their design has become so complicated that the computers themselves are needed to design their successors. They can also be conditioned to learn from their performance and improve on their design. When it becomes necessary to set up stations in places which are absolutely impossible for human habitation, either because of distance or physical conditions, computers will have to be made self-repairing, even self-making, from local raw materials. Whole races of them can be envisaged, peopling the very distant planets or even suspended in open space.

The real problem that has to be faced is a rational balance between the mental processes, enormously elaborate and multiplex that have already been evolved in the human brain, and the far cruder but also far faster processes that go on in computers. The best balance has to be struck as to which should deal with which mental task. Whatever happens, however, it is evident that we are entering into a new realm of liberty which will extend to far greater lengths than any before, because it extends outside the frame of organic evolution. It may even be, as I have discussed elsewhere,[6.19.280] that with the new possibilities we are reaching up to a new step in cosmic evolution away from the individual organisms or the society of such organisms towards an organismal electronic complex which will transcend it and may ultimately make its organismal originator superfluous.

THE NEW CHEMISTRY

The outlines of a chemistry based on the structure of the atom have already been drawn. The task of the immediate future is to make this chemistry quantitative and practically utilizable, so that science can point the way to experience and not follow it. Already, even with some empirical methods, chemistry is acquiring a capacity for making substances to order, more particularly in the field of polymers. The greatest value of the new refinements will be in the biological field, as we shall see, but sufficient can be done in the field of industrial materials to produce a major revolution. Synthetic fibres, detergents, paints, absorptive resins are already achieved examples of what chemistry can do in imitating and improving on natural products. The next stage is the synthesis of materials on the basis of theory, so that they have desirable properties not hitherto found in Nature.

Chemistry, as we have seen, has all through its history been linked to industry, or rather to the entire range of domestic, agricultural, and industrial processes requiring the production or alteration of materials. The relation, however, has been a casual one. From now on the whole range can be consciously planned and fitted together into one flow-sheet. It is only in this way that the limited resources of the earth can meet the ever-increasing needs of an industrial and scientific civilization. The emphasis everywhere will be on economy and conservation. Materials will be used not merely because they are on hand but because they are the best for the job. Atoms and molecules will no longer simply be taken up and thrown away, but will be called on to serve one purpose after another in an endless cycle. An absolute minimum, and that only of the more abundant elements, will be immobilized in structures or dissipated in air, water, or ground. The precious, sun-synthesized sugars embodied in timber will first be utilized to the full as plywood, bonded sawdust, or paper, and when they have served their turn be used as food for animals, either directly or converted by yeasts or fungi.

The immediate prospects that are offered by the physical sciences are those of complete command over the region of experience with which

266. Further uses of plastics (see plate 255) are being developed. A fibre-glass spire for the restoration of the top of St Augustine's tower in the City of London. The Church was destroyed in the Second World War and the tower is now all that remains.

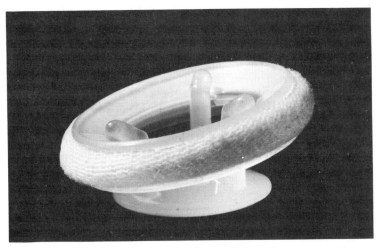

267. Plastics are also being developed and increasingly used as permanent fixtures in the human body. Photograph of a Hammersmith mitral valve prosthesis for heart operations manufactured under licence from the Ministry of Health.

we are already familiar in the ordinary operations of Nature, that is scientifically over all extra-nuclear phenomena. Before the end of the century atoms and molecules should be as manipulable as were levers, cog-wheels, and cylinders in the nineteenth century. The tasks for the next stage are to enlarge the boundaries of the physical sciences on one side towards a deeper knowledge of the interior of the nucleus and of elementary particles, on the other to an explanation of complex chemical and biological phenomena.

THE UTILIZATION OF THE PHYSICAL SCIENCES

If the sciences were developed solely in accordance with their intrinsic interests or even for the greatest assistance they could bring to human welfare, it would be relatively easy to foresee, at least in the short run, the direction of their advance. However, in the world as it is what is discovered and what is used, in the physical as in the other sciences, is even more a social and political question than a scientific and technical one.

What happens will depend on the degree with which, in different parts of the world, the new impulses to develop and use science are thwarted or distorted by economic restrictions and by military demands. If an atomic world war is avoided – and if it is not it is hardly worth while writing about the future of physics – the next few years should demonstrate the relative values of the socialist and capitalist systems in achieving scientific and technical advance and in increasing the general standard of life. This competition will be played out on a world stage, with the peoples of every country, including those of the underdeveloped countries, comprising half the world's population, looking on to know which system to choose. The major capitalist countries start with initial advantages of wealth and power, but they are hampered by their emphasis on military preparations, and their unstable economic system. The socialist countries, striving to overtake them, have had to start from a lower economic level and are weakened by their need for capital goods at the expense of consumption. Success will come to the system that can best utilize and develop science, and here, both in theory and practice, the advantage should be on the socialist side (pp. 1184 ff.).

## 10.12 Science and Ideas in an Age of Transition

The prestige of the physical sciences, great and ambiguous as it is, rests in these days on their practical manifestations in peace and war. For the first time it is evident to all that science consciously directed, rather than left to grow by blind chance, can transform almost without limit the material basis of life. Such a marked accession of human power cannot be without influence on man's ideas of the universe and his place in it – the old broad field of philosophy, now so pedantically reduced to discussions on grammar (pp. 1160 f.). Great transformations inside physical science itself have accompanied the technological and political revolutions of our time. It is no coincidence that this should be so, though it would be foolish to try to link these changes by any simple chain of cause and effect. The relations of these factors are likely to be even more complex than they were in the last great revolution of science of the seventeenth century (pp. 489 f.).

Living as we do in an age of transition, we can now see clearly enough the end of the old pattern of physical thought inherited from Galileo. This was broken at the beginning of the century by men still living, though it is only now that the full realization of the change has come through to us. What we can see only dimly is the pattern of the new physical universe that is to replace the old. We have had our

Copernican revolution, but not our Newtonian. It is not just that we live in an age of uncertainty and doubt. If the foundations of physics are discovered to be faulty they are kept well underpinned by *ad hoc* assumptions and the construction of the upper stories goes on merrily. It is rather that the flow of new knowledge has come too quickly and in too great a confusion and contradiction to be assimilated as a whole. Yet every physicist confidently expects that it will all be tidied up in time, though most feel that the process is overdue.

POSITIVISM AND PHYSICS

At the moment, as we have seen, the breakdown of the simple mechanical picture has opened the flood-gates to the wildest and most obscurantist speculation. If sense fails, nonsense, unholy or holy, may be right after all. With the majority of scientists the difficulties are not embraced but evaded by sticking very closely to the observations, even to the extent of doubting what they are observations of. The predominant philosophy of physical science in capitalist countries is positivism, an even more diluted form of the agnosticism of the nineteenth-century compromisers. Positivism is not at root a philosophy derived from physics – its politico-social origins will be discussed later (pp. 1160 f.) – but it has bitten very deep into physics, especially in Britain and America, where a traditional distrust of all philosophy makes scientists unconsciously an easy prey to the first mystical nonsense that is sold to them.

The relativitism of Einstein, the indeterminacy of Heisenberg, the complementarity of Bohr, take a positivist form, not for any intrinsically physical reason but because they were conceived by men brought up to have a positivist outlook. As the years go on these considerations seem to become less and less relevant to the present difficulties. The reason is because these have now got far beyond even mystical solutions and the world of modern physics seems so remote from any human experience or even imagination. Nevertheless, science proceeds to cope with these considerations by means of frank analogies, methods of computation which have worked for the more familiar parts and can be expected to give some coherence to the others (pp. 766 ff.).

As it stands, the whole of modern theoretical physics has no coherence: it is full of logical inconsistencies and circular arguments.

THE CRISIS IN PHYSICS AND ITS RESOLUTION

The 'crisis in physics', discussed years ago by Christopher Caudwell (1907–37),[6.42] is now officially admitted on all sides. Caudwell, in view of his lack of scientific training, could hardly be expected to appreciate the technical niceties of the difficulties that faced the physicist. The fact that he hit the nail on the head so often showed that the real problems lay as much in society as in physics. Indeed, as we have seen, the Galilean–Newtonian mechanical and atomistic picture fitted well into the whole individualistic, competitive, economic set-up of capitalism. It began to break down under its own weight of new experiments and observations, which pointed to an unacceptable interconnection between different aspects of the physical world. At the same time the very success of capitalist production, the growth of large enterprises, the concentration on empire and war, was making the capitalist system increasingly unstable.

In society, as in physics, the resolution of these difficulties was to come from excluded and neglected parts of the system itself: in politics from the industrial workers; in physics from the accumulation of long-rejected quantized phenomena – electric discharges, photo-electricity – which could not be fitted into the system. In both cases the new elements could not effectively be incorporated without radically transforming the system. These analogies, though significant enough to show some real connexion, must not be taken too literally. The content of the new knowledge of the physical world remains independent of the form of the ideas used to discover it. The expression of the knowledge may be deeply tinged with these ideas; they may, and indeed already do, form barriers to further discovery, but they do not invalidate either the experimental or the theoretical achievements of physical science.

THE CONDITIONS OF A NEW SYNTHESIS

No one who knows what the difficulties are now believes that the crisis of physics is likely to be resolved by any simple trick or small modification of existing theories. Something radical is needed, and it will have to go far wider than physics. A new world outlook is being forged, but much experience and argument will be needed before it can take a definitive form. It must be coherent, it must include and illuminate the new knowledge of fundamental particles and their complex fields, it must resolve the paradoxes of wave and particle, it must make the world inside the nucleus and the wide spaces of the universe equally intelligible. It must have a different dimension from all previous world views, and include in itself an explanation of development and the origin of new things.

In this it will fall naturally in line with the converging tendencies of the biological and social sciences in which a regular pattern blends with their evolutionary history. It will also accord with the attitude of a more integrated, that is of a socialist, society. For all these reasons the new scheme of physical science can no longer, even when it is worked out, be thought of as a final view. It will, when it has served its turn, become immersed in new contradictions and give way to a better one. Our business, however, is not to pursue such distant prospects but to cope more adequately with our present difficulties.

It is here that we must leave the field of the physical sciences to consider the other major fields of the biological and social sciences. The disturbances and controversies that have shaken science have not been confined to physics. Indeed in physics they have still preserved a more academic character than in sciences that touch closer men's individual and social lives. Nevertheless it remains a fact that the twentieth-century revolution in physics, unfinished as it is, has already profoundly influenced our knowledge of living matter. Biology can never be a branch of physics, but the new physical concepts of atoms and quanta furnish an invaluable key for opening ways to the study of organisms. They have been, as we shall see, a major, though by no means the only, factor in furthering a transformation of biology hardly less wide in range than that of physics itself.

**Table 6**

**The Physical Sciences in the Twentieth Century (Chapter 10)**

In our time it is no longer easy to present the progress of science in one table (next page). I have chosen to divide in into two – one for the physical and one for the biological sciences. This may obscure to some extent the relations between them, but these are brought out in the text (pp. 828, 871–5) and in Table 8. I have not attempted to cover the social sciences. Owing to the enormous activity of science even in this short period, it is impossible to present more than a few of the outstanding discoveries and applications of science. The columns have been arranged as far as possible to indicate interrelationships, but difficulties arise in relation to the fifth column, engineering, which takes its place because of its close connexion with electricity. This, however, breaks the close relationships between columns 2, 4, and 6, all concerned with the new knowledge of the structure of the atom. However, the position of column 6 is dictated by the close relation of the structure of matter, with which it deals, to the development of the chemical industry, with its emphasis on synthesis and plastics.

| | Historical Events | Mathematical Physics | Nuclear Physics | Electronics |
|---|---|---|---|---|
| 1890 | | | | *Crookes* cathode ray *Stoney* the electron *Lenard* positive rays |
| | Colonial wars Growth of monopolies | *Lorentz* electron theory | | *Röntgen* X-rays |
| 1900 | | *Stephan* law of radiation | *Becquerel* radioactivity | |
| | Russo-Japanese War First Russian Revolution | *Planck* quantum theory *Einstein* special relativity | *Curie* radium *Rutherford, Soddy* radioactive transformation | *J. J. Thomson* mass of electron *Langevin, Millikan,* charge of electron Electronic valve radio |
| 1910 | Increasing inter-imperial tension | equivalence of mass and energy | *Soddy* isotopes | telephony |
| | | | *Aston* mass spectrograph *Rutherford and Bohr* the nuclear atom | *Lauu* X-ray diffraction *Braggs* structure of crystals *Moseley* X-ray spectra |
| | First World War Russian Revolution | *Einstein* general relativity Explanation of gravitation | | |
| 1920 | | | | |
| | Post-war depression Fascism in Italy | *Bohr* theory of spectra *De Broglie* new *Heisenberg* quantum | First nuclear disintegration | Radio broadcasting |
| | General Strike in Britain | *Schroedinger* theory *Dirac* wave mechanics | Cosmic rays *Cockcroft, Walton* artificial disintegration | |
| 1930 | Great Depression | *Dirac* electron theory | *Chadwick* neutron *Joliot* artificial radioactivity | *Appleton* radio echoes ionosphere |
| | Rise of Nazism | *Yukawa* meson theory | *Anderson* positive electron and meson | |
| | Spanish Civil War | The expanding universe | *Bethe* nuclear origin of sun's heat | Development of radar |
| | Second World War | *Bohr* nuclear drop theory | *Hahn* nuclear fission | Servo-mechanism and electronic computers |
| 1940 | | Meson field theory Shell theory of nucleus | First fission pile | Cybernetics |
| | Invasion of Soviet Union Liberation Cold War | | **Atomic bomb** | |
| | People's Republic of China | *Dirac* quantum electrodynamics | Cosmic ray disintegrations | Electron microscope Television Radio astronomy |
| 1950 | Korean War | *Einstein* unified field theory | Thermonuclear re-actions | |
| | | | **Hydrogen bomb** | Transistors |
| 1955 | | | Synchrotron | |
| | Suez Hungary | | Anti-proton, neutrino *Cerenkov* effect *Mössbauer* effect Multiplication of elementary particles | Ocean rift system Continental drift Exosphere Van Allen belt |
| 1960 | Liberation of Africa Congo Cuba | *Lee, Yang* non-conservation of parity | | Laser Maser |
| 1965 | | *Salam, Gell-Mann et al.* principle of unitary symmetry | $\Omega$ minus baryon | |

| Engineering | Structure of Matter | Chemistry |
| --- | --- | --- |
| Development of automobiles | | |
| Steel frames and reinforced concrete in building | Measurements of increased precision of mechanical properties of materials | Synthesis of dye-stuffs and drugs |
| *Wright's* first aeroplane | | Contact catalysis for sulphuric acid |
| Rapid development Cheap automobiles | | |
| **Mass-production** | | |
| Tanks, lorries, and planes, beginning of mechanized warfare | *Braggs* structures and properties of solids | *Haber* nitrogen from air |
| *Freysinnet* pre-stressed concrete | *Kossel, Lewis, Langmuir* electronic theory of chemistry *Heitler, London* homopolar forces | Development of cellulose plastics, rayon |
| Mechanical transport replacing older forms | *Goldschmidt* geochemistry | *Fischer, Tropsch, Bergius* petrol from coal |
| Tractors, combines, and further mechanization of agriculture Precision engineering | Structure of fibres Plasticity of metals *Taylor* dislocations in crystals *Bowden* studies in friction *Orowan* phenomena of metal plasticity | Catalytic cracking petrol from crude oil |
| Development of jet planes and rockets | | Polymerization Artificial rubber Nylon and many varieties of new plastics |
| Instrumental control of production, first automatic factories | *Mott, Frank, Read* dislocation theory of plasticity and growth of crystals | Use of tracer elements in chemistry |
| **Atomic power stations** | | Isotactic polymers Photoflash spectroscopy *Semenov* kinetics of chemical reactions |
| **Automation** | | |
| Hovercraft Space Age First sputnik | *Bardeen* theory of superconductivity | Magnetic resonance methods |
| Man in orbit | Whiskers, super-strong materials | |
| Rockets supersede military planes | | |

# Chapter 11

# The Biological Sciences in the Twentieth Century

## 11.0 Introduction

To give an adequate yet brief account of the influence of the biological sciences in the twentieth century is a far more difficult task than that attempted for the physical sciences. Yet it is essential to discuss them, for it is in the twentieth century that biology, as a working and usable science, first begins to come into its own, and has already achieved decisive successes. Hence, to leave out biology would give a picture of science completely out of balance. To do it justice, however, would require the hand of one bred to and experienced in many biological disciplines, and to that I can lay no claim. Though nothing can replace direct contact with a subject, there are a sufficient number of main trends in biology which are known to others than specialists to make it possible to give a picture in at least a rough outline. Biology now touches the physical sciences at so many points that it would be difficult for anyone who has worked in the latter not to have had some practical contacts with biological topics.

In my case the relation has been closer than the average because I have, through my work in crystal structure analysis, kept in close practical contact with biological problems and have even, on questions of vitamins and hormones, proteins and viruses, among others, made some additions to biological knowledge. Besides, since my first acquaintance with the brilliant group of biochemists that gathered round Gowland Hopkins (1861–1947) at Cambridge more than thirty years ago, I have enjoyed the society of biologists, have listened to their disputes, and occasionally added to the confusion by contributions of my own.[6·157−160] This section may stand, therefore, as a record of how biology, with its social and economic influences, can appear today to a scientist working outside of, but close to, many of its disciplines.

Although the scientific effort going into biology has been far less than into physics in the twentieth century, it has led to discoveries even more important, not only because of their effect on our lives in the creation of a new medicine and a new science of nutrition, but also in our thought about the nature of life.

The biochemical revolution culminating in the early sixties, in the working out of the genetic mechanism and the relations of nucleic acids and protein formation, is a break-through in knowledge, less deep perhaps, but far more complicated and far-reaching than the discoveries of the nuclear atom which took place in the first decade of the century.

Over most of the century, however, the advances in biology were not so concentrated, although they have been on a wider front. Certainly, biology occupies a far more important place in our general life and thought than it did at the beginning of the century, but the new discoveries ensure that it is likely to take a place of even greater importance in the rest of the century.

At the beginning of the century it seemed as if the very complex and fluid nature of living things precluded their study by the same rigorous methods as had been so successful in the physical sciences. The character of biological knowledge seemed to be more primitive and qualitative – to resemble that of chemistry in the eighteenth century. This apparent lag, itself largely the product of the intrinsic complexity in biological processes, has now been almost removed. The same rigorous methods can and are being applied in biology as in the exact sciences and an increasing number of the most brilliant scientific workers have even left the fields of physics and chemistry to study the new biology.

At the same time it is becoming evident that the same degree of complexity of even the simplest forms of life is something of an entirely different order from that dealt with by physics or chemistry. What we had admired before in the external aspects of life, in the symmetry and beauty of plants and flowers, or in the form and motion of the higher organisms, now appear, in the light of our wider knowledge, relatively superficial expressions of a far greater internal complexity. That internal complexity is itself a consequence of the long evolutionary history through which living organisms have raised themselves to their present state.

The problems of biology are not simply those of the chemistry and physics of complex systems; they are not even those of chemistry and physics with something different added. At last we are beginning to see why they remain of their own kind, to be tackled by observational and experimental science in which qualitative and quantitative aspects have both to be taken into account. The very successes of physics and chemistry have ensured that biology should now present the key problems of the whole of natural science, offering a challenge to the understanding of the world in which we live, which will call for far more extensive and at the same time better co-ordinated efforts than all those which science has dealt with in the past.

BIOLOGY AS CONSCIOUS CONTROL OF LIVING ENVIRONMENT

The situation of biology in the twentieth century offers some analogy to the situation of chemistry in the nineteenth. There, as we have seen (p. 624), under the impulse of growing demands from industry, particularly the textile industry, chemistry transformed itself from a compendium of traditional recipes, ornamented rather than accounted for by a highly mystical phlogiston theory, to a practical quantitative discipline backed by a coherent and mathematical atomic theory. The exploitation and control of the living environment, always an essential task of man, were in earlier times a matter of traditional practices, each with its own language and rules, which were essentially qualitative or, where quantitative, simply based on experience. It is only now beginning to become scientific and quantitative in theory and practice.

It has been forced to do so because in the early twentieth century, largely as a result of the spread of imperialism, new industries connected with agriculture, food, and drugs have grown up which require for their efficient operations a reproducible control of biological processes and products. At the same time old traditional industries, such as brewing and baking, are acquiring an increasingly scientific biological foundation. Finally, an enhanced concern, for economic and military reasons, with the health and efficiency of workers, peasants, and soldiers has given an enormous impetus to the study of medicine. As a result biology is beginning to acquire a solid economic basis. More money is going into it and more people can afford to work on it. These very incentives carry with them the demand for higher standards of performance. The severe control that is imposed by the demand that a science must work and pay, which has made physics and chemistry what they are, is now being increasingly applied in biology. Each advance is fixed and consolidated by incorporation into some new farm implement or drug and can then form the basis for a further advance (pp. 1231 f.).

Actually these new advances of biology have arrived only just in time; for unless man acquires a better biological control of his environment the dangers brought about by his progressive destruction of the soil, combined with an increase in population, will bring back the old spectre of famine just as surely as neglect of elementary biology in the nineteenth century would have led to the return of the spectre of plague. Agriculture, from being the major human traditional occupation, is rapidly being transformed, beginning with the wealthier countries of Europe and America, into an industry which is becoming more and more scientific in character; while medicine, from being the exclusive province of doctors, proud of their traditional lore of healing, is turning into an attempt at a scientific control of human conditions, so that health and not disease will be its chief concern in the future.

LINKS WITH ECONOMIC DEVELOPMENT

The human needs that have given rise to the advance of biology, and the effects of that advance on human health, food supply, and population, involve in their interaction the most important economic, social, and political movements. We now know enough to see how the world needs to be organized so as to provide a continuously improving biological environment for all the people living in it. Nevertheless as yet only the socialist third of the world is moving in this direction. The other two thirds are still under the dominance of the law of profit. This results, it is true, in a relatively high standard of living for the most favoured industrial workers and an undreamed-of luxury for the directing few and their dependants. But for the rest, especially the 2,000 million in the colonial and 'free' tropical countries, the result is increasing degradation. Lands are neglected and people are half starved and riddled by disease because it would not pay to improve their condition. Indeed it is because of their very misery that the raw materials on which the privileged industrial countries thrive can be obtained so cheaply.

Biological science has been invoked only when these conditions have become so bad that they interfered with profit-making itself, as silicosis has in the Rand mines or malaria in the rubber estates of Malaya. For the most part, oppressive systems of land tenure and taxation, without the relief formerly provided by periodic rebellions, lack of capital, and outright robbery of the best lands by European planters, have kept down the standard of life of native populations over most of the tropical and sub-tropical parts of the world.[6.171; 6.176]

The effect of applying the bare minimum of scientific knowledge to combat disease in such countries, without changing the pattern of exploitation, has had the result of allowing the population to increase and has thus provoked further deterioration of the standard of life and exhaustion of natural resources. Equally essential applications of science to food production and soil conservation have been ludicrously small in relation to the real needs of the people.[6.174] The demands on biological science have therefore been far less than they might have been, and what has been found out has been applied only to a most limited extent. Nevertheless, these demands have produced a rapidly increasing body of knowledge, and are transforming man's potential capacity to control his biological environment.

It is this new concern with increasing yields of food and industrial raw materials for greater industrial efficiency, and for the health of the labour forces on which the whole effort depends, that determined the new character of twentieth-century biology. In essence it started before the century opened with the first burst of imperialism in the eighties. It is no accident that Manson (1844–1922), the father of tropical medicine, was a protégé of Joseph Chamberlain; or that the first large-scale drives against yellow fever started in the Spanish-American War of 1897 and that their success made possible the cutting of the Panama Canal.

In biology, it is true, there was no discontinuity at the beginning of the century similar to that which marked the emergence of the new physics. Nevertheless, it is still useful to talk of twentieth-century biology, for it is only at its outset that the first large-scale successes of the new biology were scored – the medical achievements that first made the tropics relatively safe and the plant-breeding experiments, leading to the introduction of such varieties as Marquis wheat, which resulted in widely extending the area of cultivation in Canada.

CONTRIBUTIONS FROM THE PHYSICAL SCIENCES

With the operation of these economic factors, which increased the over-all need for biology, further advances were made possible at about the same time by new contributions, first from chemistry, then from physics. The new understanding of the behaviour of the smallest units of matter, the atoms and molecules, and the techniques for studying them, had already proved invaluable in biology in the early twentieth century. This did not mean, as some are apt to think, that biology had become a branch of physics and chemistry. On the contrary, the use of physical or chemical knowledge to explain the mechanical, electrical, or chemical aspects of living organisms brought into more relief their biological aspects. These phenomena, however well they could be described in physical terms, do not occur in mechanisms made by some divine craftsman from ideal models laid down from all eternity, but in self-regulating and self-producing entities whose present form was the result of an evolution stretching back over billions of years.

### EXPERIMENTAL BIOLOGY

The infiltration of chemistry and physics into biology was not limited to the creation of the two new sciences of *biochemistry* and *biophysics*. It had a profound influence on all other aspects of biology, particularly in giving a new character and importance to experiment. The experimental method is not new to biology. It has, as we have seen, accompanied biology, especially physiology, from the days of Galen, if not earlier. Even quantitative experiment, as Borelli and Sanctorius showed (p. 474), has a long history in biology.

However, there is still some sense in maintaining that from the last decades of the nineteenth century onwards the method of experiment, first casual and limited to a few disciplines, was turning into something new; it was becoming systematic and critical.

This was the more apparent because under the influence of Darwinism the main interest of biologists had been to establish the evolutionary origin of each part of every organism by the collation of numerous meticulous observations and dissections, rather than by determining by experiment how it lived and precisely how it had grown to be what it was (p. 644). Many biologists maintained that organic Nature was too casual and unreliable to be deliberately and quantitatively varied in controlled experiments. Yet in the twentieth century just such experiments were attempted and began to yield results.

The creation of a fully experimental biology could not have occurred without the concurrence of three major factors, which could only come into action in the twentieth century. In the first place, no biological experiment of any complexity could have been undertaken or yielded significant results without being based on the enormous accumulation of observational and classificatory work in zoology and botany, mostly carried out in the nineteenth century. It was essential that the various biological experimenters should be sure, as the results of the systematists' labours to describe species unequivocally, that they were studying the same species of animal or plant. It was equally important that the anatomy or morphology of the parts to be experimented on should be adequately and reliably described, to ensure that there was nothing anomalous about them.

The second factor was the development of experimental technique in chemistry and physics, without which neither the instruments nor the reagents would have been available for biological experiments. The twentieth-century advances of biochemistry depended largely on the progress in the nineteenth of the practice and theory of organic chemistry,

The third factor was the presence for the first time of a medicine, of an agriculture, and of a biological industry sufficiently developed to demand and enable use to be made of biological experimentation. From these roots has appeared a multifarious growth of biological experiment, ranging from the statistical control of field crops to the modification of the performance of the parasites of bacteria. In it all we are beginning to see the possibility of a control of life as positive and quantitative as has already been achieved in the control of non-living matter.

NEW TOOLS IN BIOLOGY

The progress of biology has always depended, and never more so than now, on the perfection of instruments of observation and control. Until very recently these were not developed from the immediate needs of biology but were, so to speak, gifts from outside, as was the microscope in the seventeenth century. The most recent and most powerful adjuncts to biological study have also come from physics: the valve amplifier to measure the minute currents and potentials in living systems; the electron miscroscope (pp. 785 f.), which bridges the gap between the light microscope and the interàtomic dimensions studied by X-rays; and the use of isotopes and tracer elements (p. 763) which promises a new interpretation of the actual process of transformation of chemicals in living systems. Finally, the techniques of pure mathematics, especially of statistical theory, and the use of computers in applying them, have proved invaluable in extracting significant order from the characteristically irregular measurements of biological science.

Now, however, with the developments of biology itself and with the clearer understanding of the relationships between the sciences, biology is beginning to contribute to the instrumentation of the other sciences. This is partly through the need for developing, for its own use, instruments and methods which might have been, but were not, developed for immediate service to physics or chemistry. Of these, one of the most interesting is that of paper partition chromatography for which R. L. M. Synge and A. J. P. Martin were given the Nobel Prize in 1952. It is an extremely simple technique, one that needs hardly more apparatus than some blotting paper and a few solutions. Nevertheless, with the developments that have come with it, those of paper electrophoresis, radio-chromatography and gas chromatography, the analytic methods of biochemistry have reached an entirely new level of accuracy and delicacy.

268. Analysis of complex organic substances may be carried out by chromatography (see page 873). A research worker examining a paper chromatogram on
which traces of the components of a compound organic solution have separated
out on the absorbent sheet of paper. Their different colours and positions allow
them to be identified.

Indeed, the great progresses of biology of the last decade would have
been quite impossible without them. What was wanted and what was
provided was a means of analysing very small quantities almost automatically in a short time, an enormous extension of the field of analysis
which made possible, for instance, the finding of the sequences of
amino acids in proteins. Other methods are of a more purely biological
nature, such as the genetic analysis of the biochemistry of bacteria

and viruses (pp. 907 ff.), and the assay of chemical reagents or physical stimuli by their effects on organisms or physiological preparations. This is often the most sensitive method. Indeed in the time of Galvani, as we have seen, the contraction of frogs' legs provided at first the only way of detecting current electricity. Organisms can, indeed, be treated now as pieces of apparatus. Even quite complex organisms such as the lower mammals can by breeding be given a consistency of response equal to that of very good physical apparatus. By now the biological sciences have advanced so far in observational and experimental techniques that they can themselves take the lead in developing their own methods and instruments.

## THE CHARACTER OF TWENTIETH-CENTURY BIOLOGY

Until the second half of the twentieth century, the advance of biological science had been, nevertheless, a relatively confused and groping one. Even before that time, however, notable advances were made and could immediately be put to use in the practical field, notably the discoveries of antibiotics and hormones. The phenomena studied in biology are so varied and so complex and the organization for studying them was so casual that progress in biology in the twentieth century has been one of a continuous interaction of the advances in different fields, some of which will be treated in this chapter (pp. 981 ff.).

Even in the twentieth century the advance of biology continued to be held up by the old obstacles to progress that the physical sciences met and overcame in the seventeenth and eighteenth centuries – the vested interests of ignorance rallying under the flag of piety and tradition. Biology is still deeply involved in the clearing up of concepts derived from the magical age. It lies too close to our personal and social interests, and to the very structure and functioning of our own bodies, to be even as free from human passions and the effect of social forms as were the physics and chemistry of an earlier age. We have seen how in earlier periods these apparently remote subjects were the battlegrounds of violent controversies. Biology is that today; only one great battle, that for the existence of evolution, has been won, but the other battles to establish how evolution comes about and how life started on this earth have still to be fought.

The basic biological questions – those of genetics, agriculture and food supply, and of the human population in the era of the so-called population explosion, in relation to improved medical practice and the control of disease – are essentially political questions and all involve different attitudes to biological problems. Biology is also involved in questions of vital military importance – in the lawfulness of weapons of mass destruction, particularly nuclear weapons and their accompanying radioactive fall-out. These are the most vital questions in the world today and all bear on biology. It is not surprising that biology remains a chaotic subject, but great simplifying generalizations are at last appearing.

## 11.1 The Response of Biology to Social Influences

The approaches to modern biology have been along several different lines. The interest in systematic zoology and botany which, as a result of the Darwinian controversy, was the dominant one in the nineteenth century, still continues, but contributes relatively much less to the advances of the subject. Three other influences have become much more potent: those from medicine, from agriculture, and from the new biological industry. Many of the discoveries and, even more, the changes of outlook that have made of mid twentieth-century biology a radically new subject are derived from attempts to satisfy the needs of practice.[6.194; 6.223]

### MEDICINE

In fundamental biology the influence of medicine has been paramount. It is only in this century that the influence of science on medical practice, derived from the nineteenth-century pioneer work of Pasteur and Claude Bernard, began to make itself felt on a large scale. Medicine has become dependent for its supplies on important chemical and instrument industries, while in relation to its patients it has become more and more involved with organs of State power. Pharmacy, from being the collection of simples or the compounding of drastic mineral salts, has become a scientific industry, and one of no small importance even from the purely commercial point of view.

With that great achievement of the twentieth century – the development of antibiotics, both of synthetic, such as the sulphonamides, and of natural origin, such as penicillin – pharmacy has come to exert a positive effect on the whole progress of biological science, turning it in the direction of the understanding of the chemical processes underlying life. The differences between its present influence and that which, as we have seen, has been exerted by the need for finding and preparing

269. Manufacture of penicillin and other antibiotics in bulk necessitates large-scale chemical engineering coupled with the most stringent sterile conditions. A battery of antibiotic fermentation tanks, connected by pipeline to other sections of the plant, thus permitting the exclusion of undesirable micro-organisms from the entire system.

drugs in the past, are those of increased scale and efficacy. We are still very far from a rational pharmacology in which not only the apparent efficacy of the drug but also its precise biochemical mode of operation is known. Only then will it be possible to control scientifically the body processes, to restore and retain health. On the other hand, we have left behind once and for all the old philosophical or magical justifications for drugs which dominated medicine and misled science for so many centuries.

NUTRITION

In the early twentieth century a relatively neglected aspect of medicine – that of dietetics – leapt into prominence as the science of nutrition. Its study was to lead to a major scientific discovery – that of the accessory food factors: the vitamins. With this came the knowledge of how much and what kinds of food people needed to eat to keep healthy or even alive. This was the basis of the nutrition surveys and nutrition campaigns of the great depression. The work of pioneers like McGonigle, Le Gros Clark, and Boyd Orr led to the establishment of minimum standards, like that of the League of Nations in 1936 and of the Advisory Committee on Nutrition in 1937.[6.197] Ultimately, owing to the stimulus of military preparedness and war, this knowledge was forced even on governments, who had to take action to provide the food necessary to keep up their military and industrial manpower. This in turn had a direct influence on the largest and oldest biological industry, agriculture, and on the newly established food industries.

THE FOOD INDUSTRIES

Already by the end of the nineteenth century the food for the great new urban concentrations no longer came straight from the farm to the table. Increasingly the people's food has come to depend on an industry concerned with its processing, and this industry, as time goes on, has become more and more scientific. It was driven to do so in part in the simple pursuit of profit, and in part because the scandals associated with improperly prepared and adulterated foods stirred the public conscience into introducing legislation and rigid control. The growth of the food industry has led to the beginning of a rational system of preserving and preparing food. From the factory it is spreading to the home. Artificial refrigeration, which began in cold store, has entered the kitchen, and cookery, the oldest chemical industry, is at last on the way to becoming scientific. Even though less and less cookery is done at home, what is done there will have to become more scientific, if only to save time without a resulting loss of palatability.

CONTROL OF PARASITES

Nutrition is only one of the new aspects of public health that have furnished a stimulus to biological advance. The triumph over waterborne diseases by the introduction of sanitation was a major achievement of the nineteenth century. The triumph over the even more wasting, insect-borne diseases – malaria, typhus, yellow fever, and plague – by a combination of engineering and chemical methods, is that of the twentieth, a direct consequence of the drive to more intense exploitation of colonial lands by the new imperialism. This attempt brought out, far more than the earlier one, the need for combined attack. Many biological sciences such as entomology and ecology were stimulated; indeed, some, like epidemiology and parasitology, were almost created in its service.

Clinical medicine has also had an enormous effect on biological science because of the new realization of the need to invoke it to

270. Control of parasites is a vital part of modern preventive medicine. The eradication of malaria by spraying is one of the achievements of the twentieth century.
World Health Organization team prepare to go up river to reach the Punam forest dwellers in Sarawak and carry out a spraying programme.

understand and cope with the effects of disease. Indeed the very success of science in dealing with epidemic disease has caused more emphasis to be laid on chronic states, such as rheumatism and heart disease, and on the effects of the increasing number of strains and accidents produced by a mechanized civilization. For example, the universal spread of motor transport, besides multiplying road accidents, has led to gastric diseases among professional drivers.

## MEDICINE AND WAR

The extreme case is the calamity of war, which has in this century spread death, wounds, and disease more widely than in any other. Paradoxically, the urgency in war has led to a greater scientific effort in preventive and palliative medicine than any peace had produced. It was in war that the methods of blood and serum banks were first tried out. It was for war that the great potentialities of new drugs like penicillin or insecticides like DDT were developed rapidly and used on an enormous scale. More immediately, war medicine, and particularly blood transfusion and plastic surgery, have contributed to our knowledge of the working of the human body and of its means of growth and regeneration, directly and by corresponding research on animals.

All these causes acting together are creating a new human biology, which combines and revives the old anatomy and physiology of the medical schools. Research tends to take a larger share in medical training and experience and to feed into medicine able men with a scientific outlook. Indeed we are witnessing a rapid transformation of medicine from a magical art into a scientific discipline.

## AGRICULTURE

Agriculture became in the twentieth century a powerful stimulus to biological research. The changes wrought in agriculture in the nineteenth century were primarily those of mechanization. It was a matter of finding cheaper ways or, more especially, ways involving less manpower, for doing essentially what the Neolithic farmer had done in his time. The changes in twentieth-century agriculture are still largely mechanical – the tractor is a twentieth-century innovation – but they are becoming at the same time more and more biological in nature, positively, in the direction of improvements in fertilizers and feeding-stuffs, and negatively, in the continual struggle against the forces of Nature and animate creatures, in the battle against insects, moulds, and viruses, and in the conservation of the soil against erosion and sterility. Indeed the whole new science of the soil – pedology – founded by the

271. The use of fertilizers is one of the factors that has brought about significant changes in agriculture. In the Soochow region of China, a rich aquatic fertilizer plant is required and this photograph shows a green manure nursery; plastic covers are used as a frost protection.

pioneers V. V. Dokuchaev (1846–1903) and K. D. Glinka (1867–1927) in the late nineteenth century, and still bearing its Russian origin in its terms such as podsol and chernozem, is a direct outcome of an attempt to found a scientific agriculture.

BIOLOGICAL INDUSTRY, OLD AND NEW

A third source of development for biology has been that derived directly from the biological industries, old and new. Brewing, as we have seen, has furnished some of the greatest advances in early bacteriology, and now there is a growing realization that much of the chemical industry, particularly that part which depends on the utilization of natural products, can often be dealt with almost as economically by biological means, that is by the action of bacteria, as by direct chemical action. In fact we are seeing a new type of industry growing up, one carrying out on a factory scale what is naturally carried out inside the bodies of many existing animals such as cattle or termites.

Cattle do not themselves digest the grass they eat directly; rather it serves as nourishment for a host of bacteria that inhabit their various stomachs, and they live on the soluble products of these bacteria and on their dead bodies.

In future we may find that a whole industry based on a full knowledge of bacterial and algal metabolism – a microbiological industry – producing drugs, like penicillin, food, and industrial products, will enter into effective competition with the purely chemical industries over a large range of products, particularly where it can be combined with an effective utilization of the agricultural wastes of today. Indeed in the latter part of the twentieth century there may be as big an industry based on applied biology as there was in the nineteenth century based on applied chemistry (pp. 828 f.).

## PHASES IN TWENTIETH-CENTURY BIOLOGICAL PROGRESS

These general considerations on the factors influencing the development of science in the twentieth century need to be supplemented for historical perspective by considering the relation to biology of the political and economic events of that disturbed and violent period. These have already been discussed in relation to science in general in the introduction to Part 6 and to the physical sciences in Chapter 10.

It is not easy to trace in the biological sciences any clear-cut stages of advance such as are evident in modern physics. There is no question therefore of tracing close parallels between internal and external developments. Nevertheless biology has proved most susceptible, just because of its relatively weak economic backing, to large injections of financial aid. In medicine and agriculture in particular, rapid advances have been made under the stimulus of war. Indeed at times the economic interests of the moment have given a general tinge to a biological research, as for instance, in nutritional biochemistry in the thirties, or antibiotics during the Second World War.

The major divisions of the history of our times are deeply etched by bitter experience into every scientist's, indeed into every adult's, mind. The two great wars and the depression that fell between them suffice to divide up the fifty years into five periods of unequal length.

In the first period, up to 1914, the sunset of the liberal age, biology flourished in the wake of expanding imperialism. It was the period of the first great triumphs of medicine against malaria and yellow fever, and marked the new turn in animal- and plant-breeding beginning to pay dividends in Australia and Canada.

272. The utilization of biological science for destruction was seen first in the poison gas attacks in the First World War. The gas attacks were often preceded by smoke screens that themselves acted as irritants and reduced fighting efficiency. Photographed near Sedan in the Ardennes, May 1917.

### BIOLOGY IN THE FIRST WORLD WAR

The First World War was an interlude, which, except in America, distracted biologists from their research. It did, however, serve to show that anti-epidemic measures were by then adequate, for the first time in history, to preserve huge armies indefinitely in the field, though they failed to check the subsequent epidemic of influenza among half-starved civilians which killed many millions more than the battles. It also gave a foretaste of biological warfare in the form of poison gas. This first clear application of modern science for destruction provoked such a reaction among scientists and the people that despite the cease-less official research on it in inter-war years no belligerent dared use it in the Second World War. Gas had, in fact, been used once more by Mussolini in his civilizing mission against the Ethiopians, against whom, being black, and unable to retaliate, the use of any means of warfare was deemed legitimate.

### BIOLOGY IN THE INTER-WAR PERIOD

The inter-war years comprise first the aftermath of boom, slump, and boom again, then the great depression of the thirties, and finally the rise of Nazism and the drift to war. At first it was the stimulus of starvation and disease that concentrated biological attention on nutrition and counter-epidemic research. This gave a great impetus to the use of the earlier discovered vitamins and their related hormones. The first post-war years are most aptly characterized by the coming of age of *biochemistry*.

The depression, with its picture of poverty in the midst of plenty – of coffee burned, crops ploughed in, and millions of skilled workers unemployed – showed rather the futility and frustration of biology in the prevailing economic system. During the same period the rapid development of medicine and agriculture in the Soviet Union as part of the First Five-Year Plan began to show the existence of a working alternative.

In the later thirties the shadow of war lengthened and the violent spread of the race theories of the Nazis, with their perversion of science, recalled to biologists, and particularly to geneticists, the social implications of their work.

BIOLOGY IN THE SECOND WORLD WAR

It was, however, only in the Second World War that the full practical potentialities of biology began to be realized. The need to protect the fighting men against disease, especially in the tropical theatres of war, and the need to minimize the consequences of wounds, led to all-round advances in sanitation, medicine, and surgery. D D T, penicillin, paludrin are all essentially war products. At the same time the overriding need for food stimulated agriculture and the processing industries.

POST-WAR BIOLOGY

Some of these good effects lasted, others did not, into the confused post-war period. In one respect the military use of science, which had largely been confined to the physical sciences during the war, was itself turned to the biological sciences when peace came. The study of radioactive poisons, arising out of atom-bomb production, the experiments and trials of bacteriological weapons, seem to open a new era of biological warfare. Even the trials of the hydrogen bomb have shown its efficacy as a spreader of poison, and this has been only too tragically demonstrated on the bodies of Japanese fishermen and Pacific islanders. Only a public opinion enlightened by biologists conscious of their social responsibilities can prevent it turning into a grim reality, and endangering not only the whole human race, but the very existence of life on this planet.

Nevertheless, as in other fields, the effects of the war were not entirely negative. Through its developments in electronics it has helped the infiltration into biology of new physical techniques, radioactive tracers, ultrasonics, electron microscopes, electro-encephalographs. The second post-war period marks the coming of aid to biophysics.

The last triumphs of the last decade had, indeed, been due to the combination of biophysics with the older biochemistry in one integrated discipline using many different branches of investigation on different levels. The post-war period also witnessed a multiplication of antibiotic drugs together with the beginning of a rational approach to pharmacology.

At the same time biology has been entering more and more closely into agriculture. There has been a realization of the urgent need in many parts of the world to stop the waste of natural resources and to build up new sources of food for an ever-growing population leading to an over-all concept of biological engineering and the transformation of nature, first adopted as a conscious aim in the Soviet Union and now spreading to many parts of the world. The great task of supplying an integrated programme for dealing with the short-term and long-term problems of the under-developed countries has been envisaged but hardly as yet carried out. It is here that the economics of the Cold War have done their worst to stop effective progress and to widen even more the gap between conditions of life in the advanced industrial countries and those to which the benefits of science have not reached or are likely to reach. What is needed is complete geological, physical, and biological synthesis in which new ecologies can be substituted, not so much for original, natural ecology but for the erroneous ecology imposed by man under the drive of an essentially exploitative, profit-making economy.

GROWING POINTS IN BIOLOGY

This summary account of twentieth-century advances in biology may serve as an introduction to a more detailed study of the progress of the different biological disciplines. Enough has been said at this stage to show something of the way in which powerful economic and social forces have contributed to the rapid advance of biology in our time and of the reciprocal influence of this advance on the course of economic development. However, it is not only through the impetus given by social forces to the different branches of biology that these have affected its progress. The other part of the story is the influence of these economic and political forces on the inner workings of biological thought, on the moulding of ideas and the opening or closing of biologists' minds to different types of explanation of the phenomena, and consequently on the types of observations and experiments made.

These influences will emerge only when we come to examine, still broadly but in more detail, some of the major branches of biology that have shown the greatest and most fruitful advances in the last fifty years. Those I have chosen are: 11.2, *biochemistry*; 11.3, *molecular biology*; 11.4, *microbiology*; 11.5, *biochemistry in medicine*; 11.6, *cytology and embryology*; 11.7, *the organism as a whole and its control mechanism*; 11.8, *heredity and evolution*; 11.9, *organisms and their environment: ecology*; 11.10, *the future of biology*. The treatment can obviously not be comprehensive because my knowledge is very various over different parts of these fields. However, I have tried to give a balanced treatment as far as my abilities extend.

In dealing with each topic, containing as it does a bewildering complexity of sub-topics, it is impossible to maintain even the degree of historical treatment which was achieved in dealing with the physical sciences. Between topics the time correlations are even more difficult to follow. Seen, however, against the general historical background already sketched here and in the introduction to Part 6, many of the particular advances can help to illustrate a close or remote connexion with political or economic events.

These eight tracts of advance in biology are not separate, but continually overlap and merge into each other, besides incorporating a growing proportion of the physical sciences. Among the eight the first five are more closely connected with medicine, the last three with agriculture. Enormous advances have been made in all these branches in the twentieth century, in fact many of them are essentially twentieth-century sciences.

## 11.2 Biochemistry

The science of biochemistry is far more than the application of chemistry to biological problems. It is rather an attempt to discover and ultimately to imitate the far more delicate and controlled chemical operations that occur in living organisms. Biochemistry has grown into a separate discipline, not only because of the different field in which it operates, the chemistry of the products of life, but also because of the different methods it employs. Its object is not only to probe into the structures of the molecules that are to be found in living structures but also into their overall mode of reaction, both separately and in combination. For this purpose it has developed a number of different kinds of approach in which whole organisms or whole organs are studied intact or broken down to different degrees. This specific biochemical approach, therefore, is inward from the organism to the molecule, using as time goes on more and more refined methods of measurement, bringing in physical methods such as tracers, chemical methods such as various types of molecular separation processes (pp. 873 f.), and more purely biological methods such as genetic and immunological analysis.

The drive behind biochemistry has always been and remains a humane or a utilitarian one – to improve medicine, on the one hand, or agriculture and old industrial processes, such as brewing, on the other. It is a self-sustaining and, indeed, a self-accelerating process. Biochemistry had its origin in the study of fermentation, and its establishment as a separate science may be taken somewhat arbitrarily to date from the discovery that E. Buchner (1860-1917) made, almost accidentally, in 1897, when he found that crushed yeast could cause sugar to ferment even though no living cells were present. This showed that a dead chemical substance, what was called an enzyme – en zyme = in yeast – was responsible for fermentation and similar substances for most other chemical reactions occurring in living matter.

It was to take some forty years, however, even to begin to understand the nature of enzymes and the mechanism of their action. In the great controversy of the nineteenth century – that between Pasteur and von Liebig on the nature of fermentation – both were right and both were wrong (pp. 647 ff.). Liebig was after all justified in claiming that fermentation was caused by a chemical. On the other hand these substances were not laboratory chemicals but could be produced only by living organisms, and this justified Pasteur, who had claimed that life played an essential part in fermentation. Non-living ferments like the diastase of malt had, it is true, been known and used by man since the dawn of history. The importance of Buchner's discovery was that it proved the long suspected fact that reactions inside the cell, often attributable to mysterious vital forces, were due to intra-cellular ferments or *enzymes*.

What intrinsically marks off biochemistry from the more classical organic chemistry – itself arising from the study of products of life – is that it deals with chemical processes as they are carried on in and around

273. Frederick Gowland Hopkins (1861–1947) was a pioneer in biochemical and nutritional research. From 1930-35 he was President of the Royal Society and this portrait by Meredith Frampton shows him in typical mood surrounded by some of his scientific equipment. In his hand he holds a small direct-vision spectroscope and on his pad are results of a spectrum analysis.

the cells of living organisms by means of enzymes. For example, the two major operations carried out by largely all living organisms – fermentation and oxidation – and one other, on which today all the rest depend: the photosynthesis of green plants. All are simple in their components, but are executed in an extremely complex way through a number of steps, each operated by a specific enzyme.

It is quite impossible in the small space of this section to attempt to unravel and present the story of biochemistry as it should be presented in its historic order together with its interactions with medicine, agriculture, and industry. The starting materials are so various, including as they do a not quite arbitrary selection of a few thousand out of the many billion distinct chemical substances that could be found in living organisms. Even more varied and multiple are the reactions between them.[6.153; 6.181] The clues to this maze were, in fact, provided by the human, social, and economic selection of definite problems in the effort to explain and control useful or noxious natural processes. The need to promote or check fermentation or growth, to understand the action of drugs, to assay the real value of foods, have all played their part in the development of biochemistry and, through the successes registered at every stage (the discovery of vitamins, hormones, antibiotics), have step by step increased the prestige and activity of biochemistry. Off the main line of medical and industrial concern have been many fascinating and rewarding side-tracks, and even pure curiosity has played its part. The great Hopkins began his biochemical researches by an analysis of the pigment of butterflies' wings – a lead into the important group of pterins related to pantothenic acid, one of the constituents of vitamin $B_2$.

Even if a history of biochemistry could be compressed into a small space it could not be presented to a non-specialized reader without explanations longer than the story itself. Faced with these difficulties, and at the risk of exasperating my biochemical friends, the best I can do is to abandon the historical approach and to discuss a very limited selection of aspects of biochemistry which illustrate particularly well the interaction between scientific research and social forces. Further, in order to make them at all intelligible I will treat them in the light of my present knowledge of biochemistry, out of date though it inevitably is, with the consequence that they will necessarily appear on a quite different background of scientific knowledge than that of the times in which they were made. The order I am adopting is logical rather than historic, but even then it is difficult to make each part depend only on what has gone before, and not as well on what is to follow. It would consequently, for those who are sufficiently interested, warrant a second reading.

I am beginning with a short description of the intermediate molecular building-blocks out of which most living matter seems to be made. This is necessary to introduce a discussion of the action of enzymes and co-enzymes and of the processes of fermentation, oxidation, and photosynthesis. I then turn to the story of vitamins, trace elements, and hormones as further examples of the biochemical action of small quantities of special substances. From there I will continue with a more general discussion of metabolism and of the character of life as a thermodynamic process. This is what might be called classical biochemistry. I have deliberately separated in this division the deeper aspects of the explanation of biochemical processes in terms of molecular structure, reserving it for what is virtually the new subject of bio-molecular studies in the next section (11.3). It is here we find the antecedents and the beginning of the modern revolution in biology – an explanation in terms of molecular structure. Here the great advance has been in the explanation of the molecular structure of the information, which is built into the nucleic acids, and of its exchange and storage, and their relation to the synthesis of specific proteins. Here we have, as we shall see, the synthesis of the basic biochemical knowledge with that coming from outside biochemistry proper, in crystal analysis, on the one hand, and genetic studies, on the other.

THE BASIC MOLECULES OF LIVING ORGANISMS

Recent work has confirmed that it is the activity of continuous cycles of chemical processes rather than the existence of any material substance that gives life its specific character (pp. 902 ff.). Before, however, these processes can be discussed it is necessary to say something of the forms of the molecules that are intermediate between the simple inorganic gas molecules like ammonia or carbon dioxide and the highly complex proteins and nucleic acids that are essential to present-day organisms. Logically, and probably historically as well, smaller molecules with a dozen or so atoms come before the larger, thousand-to million-atom molecules.

All of these have indeed been shown to be decomposable into a relatively small number of types, themselves mostly falling into four major groups. There are (1) the twenty-odd *amino-acids* that make up the *proteins*; (2) a few nitrogen-containing double-bonded ring molecules, including the *purines* and *pyrimidines* of the *nucleic acids*, the *pyrroles* and *porphyrin* of cell pigments and many physiologically active

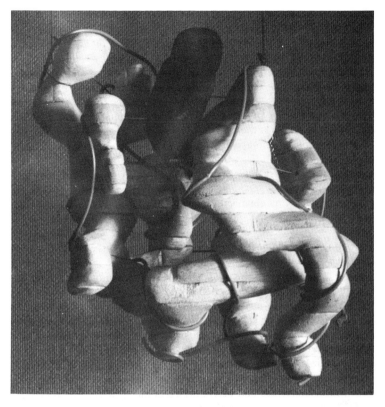

274. A model of the molecule of the enzyme lysozome from hen egg-white. The structure of the molecule was determined by X-ray studies. Chemically it is found to contain a polypeptide (protein) and structurally this is found to run through it as shown by the wire on the model.

*alkaloids*; (3) the *vegetable acids* and the *carbohydrates*, mostly *sugars* and their derivatives; (4) the *fats* and their related *sterols*. All living things on this earth the biochemistry of which has been studied seem to be built of these basic molecules; and though relatively few species have been studied, these should be a representative sample.

Of these, the amino-acids, or at least the simpler of them, seem to be the most primitive, and indeed have recently been made by Miller[6.202] from ammonia and carbon dioxide acted on by light. The nitrogen-containing ring compounds seem to be derived from the first by ring formation and dehydrogenation. The sugars and carbohydrates seem now to be produced by photosynthesis from carbon dioxide and water, but this is a complex process, and originally they may well have come from the first group by removing nitrogen. The origin of fats and sterols is only now being elucidated, but they may have appeared very early, as their analogues appear in meteorites (p. 986).

It is not only the presence of relatively restricted groups of basic molecules that reveals the common origin of present-day life, but also the presence of common paths of synthesis and breakdown in all living organisms, the plants being more prominent in the former, the animals in the latter. The fact that, barring poisons, every animal can get some nourishment from every plant and that ultimately all animals live on plants shows that biochemically life is a unity.

MODE OF ACTION OF ENZYMES

That unity is maintained by the action of linked chains of reactions now catalysed by *enzymes*, though the present enzymes cannot have been the first molecules with this role. The understanding of enzyme action, by which a small particle of even a crude preparation, like rennet or malt, can transform an enormously larger quantity of the so-called *substrate*, like milk or starch, had to wait until enzymes could be prepared in a reasonably pure state. This was not achieved until the middle twenties, and even now only a few dozen enzymes have passed the test of crystallization, though fairly pure preparations of hundreds of others are known.

Only when enzymes had been purified could their enormous efficacy be fairly judged.[6.183] One molecule of an enzyme like peroxidase can activate a million molecules of hydrogen peroxide per second. The fundamental importance of the purification was to show that a crude enzyme, the so-called zymase of yeast, did not in one stroke turn sugar into alcohol and carbon dioxide, but that the process of fermentation was carried out by some twenty separable enzymes, each responsible for one detailed chemical step – removing an atom from the substrate molecule or shifting a chemical bond. It appeared in fact that the biological transformations of chemical substances in the cell were very similar to those occurring in a modern chemical factory, where each reaction vessel carries out only one operation and passes on the transformed material for the next to deal with. Further, each separate step was found to involve a very small energy change, which ensured that the reaction could proceed at comparatively low temperatures without giving off enough heat to raise it markedly. An enzyme transformation system is like a pair of steps, which enables the reactants to get over a large energy barrier without needing the energy or the high temperature necessary to jump over it in one go.

Once the enzymes could be purified it became apparent that most of them were, or contained, proteins. It had long been known that proteins or albuminous substances, like white of egg or lean meat, were found in all living cells and, in a hardened form, in integuments, like silk, wool, or horn. Engels had already in 1877 referred to life as 'the mode of existence of proteins'. Here, for the first time in purified enzymes, there began to appear at least one reason for their importance: their capacity to promote biochemical changes. Later we shall have something more to say about the structure of the proteins. For the moment it is sufficient to say that most protein enzymes are composed of large soluble molecules of a thousand or more atoms containing both acid and alkaline groups.

### BIOCHEMICAL METHODS

It is principally around the action of enzymes that biochemical methods, as distinct from those of physical or organic chemistry, have grown up. The art of the biochemist consists of separating from a piece of mashed tissue – like liver or seed germ – the various enzymes it contains. Besides using all the techniques of chemistry, old and new, the biochemist operates with devices learned and adapted from the enzymes themselves. It is often possible by the use of certain drugs to poison or inactivate some particular enzyme and thus to stop the chain at a corresponding point and to find the intermediate product. The very activity of the enzyme measured by the rate at which it transforms the substrate may serve to track it down. A more active preparation must contain more enzyme. If further fractionation does not seem to improve the activity it is probably not far from being pure.

### THE WITCHES' CAULDRON

This method of concentration by specific activity is one of the most powerful tools which the biochemist has taken from classical chemistry – the Curies used it to isolate radium – which in turn had derived it from the practice of the miners. Using these methods, once an activity is recognized a search can be made for materials which contain it to a considerable degree, and when the best is found it can be purified, yielding often in the process associated substances of unexpected properties. The raw materials are as varied as those of the primitive medicine man or the witches of *Macbeth*:

> Fillet of a fenny snake,
> In the caldron boil and bake;
> Eye of newt, and toe of frog,
> Wool of bat, and tongue of dog,
> Adder's fork, and blind-worm's sting,
> Lizard's leg, and howlet's wing –
> For a charm of pow'rful trouble,
> Like a hell-broth boil and bubble.

Now, however, they are no longer mixed but carefully separated. It was in this way that not only enzymes but also vitamins, hormones, and antibiotics were detected and purified.

Five decades of patient work by a growing band of biochemists – in Britain there were only fifty members of the Biochemical Society in 1911 while there were over 3,500 in 1965 – has unravelled a few complete reaction chains and found some hundred enzymes and other biologically active substances. Of the latter, with smaller molecules, many have been analysed and a few synthesized by the methods of organic chemistry.

CO-ENZYMES

As the reaction chains promoted by enzymes began to be studied more carefully, it was found that the proteins in the enzymes were not acting alone. Equally necessary to the progress of the reactions was a small quantity of non-protein material, usually soluble and of small molecular weight. The first of these co-enzymes – cozymase – was detected by Harden (1865–1940) and Young in 1906, and identified as a dinucleotide of nicotinic acid, the anti-pellagra vitamin, by Elvehjem in 1937. Not as many co-enzymes are known as enzymes but the same one can act for several enzymes. The function of the co-enzymes has been found in several cases to be that of accepting and passing on atoms or small molecules released by the main enzyme reaction. Riboflavine, for instance, acts as a hydrogen donor for transforming oxygen to hydrogen peroxide.

RESPIRATORY PIGMENTS

This linking of a protein enzyme with a small but active molecule brings out the close parallelism between enzyme action and that of the so-called respiratory pigments, like the haemoglobin of the blood or cytochrome of the cell. These consist of a protein globulin loosely bound to a brightly coloured and usually metal-containing porphyrin group. This combination seems to enable a small molecule like oxygen to be held very lightly, so that it comes on and off readily. In this way the respiratory pigments serve to carry out the critical step of introducing and removing small molecules in the biochemical system.

### TRACE ELEMENTS

Their specificity depends very much on the associated metal: thus only iron will work in the vertebrate blood pigment haemoglobin, vanadium in that of the related sea squirts, copper in the blood pigment of snails. As these substances are very active and only one atom of metal is needed for a protein molecule containing 5,000 or so atoms, the amounts of metal required are very small. Without it, however, the system will not work and the animal or plant will die. This is the explanation that was found for the mysterious pining diseases of cattle and sheep feeding on meadows deficient in some one metal. Pining in cattle, for example, can now be cured by the application of 28 ounces of cobalt per acre. The use of such *trace elements* is likely in the future to extend widely the area of profitable agriculture.

275. Out of barren desert a 6 million acre region in Southern Australia is being transformed by the use of trace elements zinc and copper to make good mineral deficiencies that have prevented settlement before. Desert scrub side by side with reclaimed land shows the effectiveness of the treatment.

## PHOTOSYNTHESIS

The porphyrins are coloured molecules, that is they react to visible light. It is not surprising therefore to find one of them – chlorophyll – as the overwhelmingly most widespread and successful light-trap molecule in *photosynthesis*. Through this one molecule passes all the energy from the sun that makes plants grow, animals move, and men think. The crude product of photosynthesis in higher plants seems simple enough. Carbon dioxide is taken in from the air, reduced to carbon, combined with water to form a carbohydrate – sugar, starch, or cellulose – and the extra oxygen restored to the air.

The real process, now nearly elucidated through the work of armies of biophysicists and biochemists, in which the names of Van Niel and Calvin and Kamen are prominent, turns out to be far more complicated. It appears that in the first stage in the chloroplast the light is used to charge certain intermediate co-enzyme-like molecules electrically and that the energy stored is used to build atmospheric $CO_2$ step by step into sugar and other biological molecules, while the hydrogen needed is extracted from water leaving the oxygen to replenish the atmosphere.

The discovery of the action of respiratory pigments, enzymes, and co-enzymes pointed the way to the explanation of phenomena very long known: the violent effects of certain substances on large organisms even when administered in extremely small quantities. This knowledge, in fact, dates from the Old Stone Age, with the first discovery and use of poisons. The word *toxon* in Greek stands for arrow and poison. In a few simple cases the mode of action of poisons can be explained. Cyanide and carbon monoxide, for instance, act by combining with the haematin of haemoglobin and oxidative enzymes more firmly than the oxygen that they should carry, and thus block the main oxygen-transporting mechanism.

## THE DISCOVERY OF VITAMINS

The importance of very small quantities of chemicals in biological processes was also discovered in modern times, rather paradoxically in reverse, by the effects of not having them. In the past many diseases were attributed, and quite rightly, to deficiencies in diet. Of these perhaps the most important was scurvy, the sailors' disease. It was also the first to be recognized as a deficiency disease. Already in the eighteenth century, Captain Cook had kept his crew free of it by a permanent provision of fresh fruit. But this knowledge was not scientific, and tended to be forgotten in the nineteenth-century vogue of the germ theory of diseases. It was the genius of Hopkins[6.189] that first drew attention to the presence in full diet of small quantities of substances in the absence of which growth did not occur and degenerative symptoms appeared.

These accessory factors, later to be known as *vitamins*, gave an immediate impetus to the study of biochemistry, because here at last were chemicals that could be used, and used immediately, for curative purposes. Once the idea gained ground that a particular condition was due to a deficiency it became a matter of hard work and chemical technique to find out what the deficiency was, to isolate the substances that could cure it, to determine its formula, and finally to synthesize it. There were, of course, many difficulties; although some vitamins were simple, such as, for instance, vitamin C or ascorbic acid, first isolated by Szent-Györgyi, who defined the vitamin in a paradoxical way as 'a

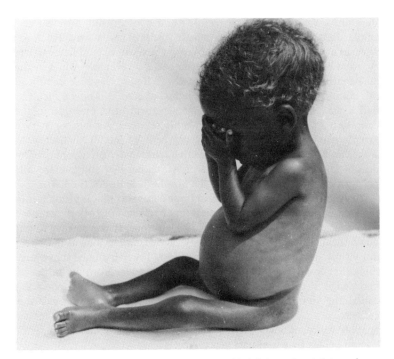

276. Malnutrition may take many forms. Protein deficiency has led to oedema, skin lesions and uncurled greying hair in the child in the photograph. Protein-rich foods can effect a cure. Photographed for the Food and Agriculture Organization of the U.N. by M. Autet.

substance that makes you ill if you don't eat it'. Other vitamins were very complicated indeed. What was first called vitamin B was found to contain at least fifteen different substances, each needed for the carrying out of some different function in the body. Many, possibly all, vitamins appear to function as co-enzymes and many represent only those that, as they normally occur in food, the organism has lost the ability to synthesize.

SOCIAL EFFECTS OF THE KNOWLEDGE OF VITAMINS

The discovery and isolation of the vitamins, and the determination of the quantities of each necessary to maintain health, provided in principle the first approximately complete and quantitative assessment of the food requirements of human beings. In the twentieth century science thus put into the hands of humanity a means of ensuring the good life, as far as food could do it, of the population of the whole world. Vitamins are fairly widely distributed and consequently a mixed and ample diet always contains enough of them. This is why deficiency diseases are primarily diseases of poverty which can be cured completely by good economics and good government. For example, while in the nineteenth century rickets, with its twisted limbs, was so common in this country as to be known as the English disease, it is now difficult to find a case. This is a very recent achievement, and one due to the operation of maternal and child health services. As late as 1931 a sample survey showed over eighty per cent of school children with some clinical sign of rickets. On the other hand under-privileged peoples do not fare so well. In large parts of Africa beriberi still exists, while pellagra is common in Italy and the Southern States of the USA.

The value of scientific research in these cases was that it brought into the light of day facts about nutrition that had previously been confused with a large number of irrelevant considerations. It was so easy to put down the illnesses of the poor to drink or vice, and as long as they were not visibly starved or dying for lack of food it was assumed that everything that could be done for them was being done. Now, with the new knowledge, it could no longer be hidden that withholding good food containing vitamins was an actual crime against humanity. Once this knowledge was well established and widespread it became no longer possible to tolerate what was effectively the crippling and maiming of human beings by social negligence.

Characteristically enough, it was not these considerations but rather those of the fitness for armies to fight in the Second World War that led to the fully efficient and official taking up of the applied science of nutrition. This was done to such good effect that it was possible to keep the British population actually healthier than they were before the war, on a very much reduced gross diet, which, in the absence of a knowledge of vitamins, would inevitably have meant a great incidence of deficiency diseases, particularly among children, and a general increase in epidemic disease as well.

## HORMONES

The importance of special molecules in very small quantities was not, however, limited to molecules taken in food. At the same time as these researches were going on, others were showing that many bodily conditions were dependent on the existence of minute quantities of substances produced inside the body itself, usually in special places: the so-called ductless glands whose function had been a mystery to the earlier anatomists. Thus a new group of substances was discovered, the *hormones* or messengers, as E. H. Starling (1866–1927) first called them in 1905, such as oestrone and its related ovarian hormones connected with the female sexual cycle and lactation. Another is thyroxin, the failure to produce which may cause goitre and cretinism. Iodine is the key element in thyroxin, and its absence in many areas is the basic cause of these diseases, which can be prevented by an adequate distribution of iodides. In other cases, such as insulin, the problem was more complicated, the hormone being itself a protein and therefore not yet synthesizable. The sufferer from diabetes is dependent on the hormone production by proxy of another organism or on insulin derived by extraction from the pancreas of cattle and sheep. Unfortunately the incidence of diabetes throughout the world is greater than the potential supply of insulin from animals. Unless we are willing to tolerate the death of hundreds of thousands from preventable causes, there should be a most determined and well-backed effort to synthesize insulin or insulin substitutes.

## PLANT HORMONES

The successes of vitamin and hormone research were not limited to animals. Went and others in 1928 began to study by biochemical means the way in which the growth of plants was affected by external stimuli such as light and gravity. To say that plants naturally grow upwards and towards the light is simply to allow familiarity to conceal ignorance. To measure how they grow is an essential step to understanding; but only by experiment, controlling and varying the state of the environment, could the process begin to be understood. In this way natural substances, the *auxins*, were discovered that produce lengthening of cells and hence growth, which may be straight or crooked, according as the auxins are evenly or unevenly distributed. Later it was found that artificial substances, not very similar chemically to the natural auxins, had similar effects. These hetero-auxins are now widely used for promoting growth, particularly the rooting of cuttings. In larger doses they produce unregulated growth and death, and are therefore beginning to find application as weed-killers. It is characteristic of the pathological state of the capitalist world that others are being developed at great expense and in deep secrecy to be used to blast enemy crops in biological warfare, and have actually been tried out, without effective protest, against peasants in Malaya by the British and in South Vietnam by the Americans.

The study of vitamins and hormones, and even more the often dramatic effects in the practice of administering them, make it very tempting to think of organisms no longer as mechanical machines but as chemical machines, whose performance is entirely determined by the totality of active agents administered to them. As experienced biologists and even biochemists point out, it does not follow that if the giving of a specific chemical produces a certain physiological result, it is the same or a very similar chemical that produces it under healthy conditions. There are many other chemical and neurological factors to consider, and the same result can be reached by very different paths. Nevertheless, this knowledge should not lead to an all-pervasive scepticism or mysticism in biology. Rightly considered, it should be a spur to deeper and more comprehensive biological research.

IMMUNOLOGY

So far we have stressed the activity of molecules in organisms. Some have another property, that of specificity, which is also peculiarly associated with proteins. Pasteur had discovered, almost by accident, in the reaction of induced immunity, how a harmless vaccine taken from a concoction of dead bacteria could immunize a patient against an attack by the same bacteria in a virulent state. This became the basis of the new science of *immunology*. Its practical successes have been registered in the virtual abolition of diseases like diphtheria.

In essence this represents only a further stage in the bringing into the light of processes that have for millions of years served to protect animals from infectious disease. Their recognition and use by man are also hidden in the mists of history. No one knows the origin of the practice of inoculation for smallpox long practised in the East, but it

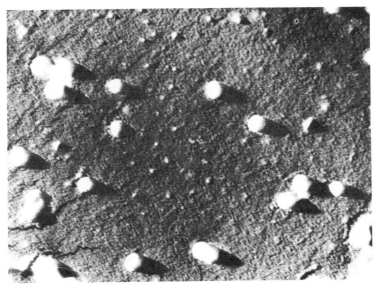

277. Electron-micrograph of influenza virus. The virus is smaller than the wave-length of light and cannot, therefore, be observed in an optical microscope. The size of the virus is of the order of a few millionths of an inch. Photograph taken for the World Health Organization.

owes little to science. It was from this, however, that Jenner in 1796 drew his practice of vaccination, which was important because it was the first scientific use of the principle of protective immunization, recognized traditionally by the milkmaids, of the milder bovine form of the disease. Nearly eighty years were to pass before this first break-through was followed up, and it was not until this century that the principle of immunity found a wide range of application. The same effect became apparent later when the old expedient of blood trans-fusion was seriously attempted in human subjects.

### BLOOD GROUPS

At first, among the successes serious accidents occurred; and it was discovered that the proteins in some people's blood were such as to react with and indeed precipitate the blood cells of other kinds of people. That was the beginning of the study of blood groups by Landsteiner, which was to prove of such inestimable value in saving lives in war and also in peace. Both these reactions depend on the fact that proteins are highly specific; that each kind of protein can act as an agent in the body to produce an antibody which will precipitate this and only this protein in future. The mechanism of this reaction is still obscure, but enough is known to show that only a particular part of the protein molecule is involved. Its further study is bound to throw light on the biologically essential details of protein structure (pp. 906 ff.).

METABOLISM

One of the central problems of biology is that of metabolism. As already mentioned, some of the processes of metabolism – the burning up of sugar, for example – have been more or less worked out; but far more remains to be done, and the study on the constructive part of metabolism or *anabolism* has hardly started. One thing, however, has become clear very recently, particularly by the use of tracer elements; this is, that both the anabolism, or building up of compounds from simple structures in the body, and the *katabolism* or breaking them down, are taking place at much greater rates than had hitherto been thought. The molecules in our bodies and in all organisms are in a perpetual state of reconstruction, and the atoms flow through them in an almost continuous stream. It is probable that none of us have more than a few of the atoms with which we started life, and that even as adults we probably change most of the material of our bodies in a matter of a few months.

BIOCHEMICAL CHARACTER OF LIFE AS A PROCESS

What is permanent, then, in an individual life is not the matter but the forms and reactions of the molecules out of which organized beings are made. The actual matter of organisms seems to be essential mainly because it is needed to execute the continual cycles of chemical changes which are life. These changes must be *more or less balanced* in every living cell, as they are in the organism as a whole. This – more or less – implies that inside each cell and in the organism as a whole the cycles are never complete; that growth or degeneration is the rule over the life-span, a distant echo of the 'generation and corruption' that ruled according to Aristotle in the sub-lunary sphere (p. 200). Further, the balance, as Claude Bernard pointed out, is within limits a stable one: the organism reacts so as to keep both its internal and external environment constant. It is only when the limits are overstepped and one type of change gets out of hand that the living cell or the organism ceases to function in a co-ordinated manner (or as we say dies). Even after this has happened many of its component parts, such as enzymes in the case of the cell, or whole cells in the case of the organism, are for the time being as effective as before.

The essential feature for any one organism, while alive, is the sequence and co-ordination of *processes* rather than any architecture of inert matter. Taken over the whole of life on this planet, the importance of process looms even larger. In reproduction as in growth, but at a much slower rate, the cycles of processes are modified. The actual processes and the structures which maintain them acquire their full meaning only when they are seen as the product of a long evolution, in the first place a chemical evolution.

The exploration of the nature of the fundamental chemical processes of living matter has begun only in the last decades and is at present in a very active phase of discovery. All these processes seem to be brought about by enzyme–co-enzyme systems. Indeed it would appear that most of the free protein molecules in cells are functioning as enzymes. The role of the co-enzymes, and particularly of the phosphorus containing nucleotides – the components of nucleic acid – seems to be of key importance. They seem to be the link between the energy-releasing, katabolic processes and the energy-absorbing and structure-building, anabolic processes.[6.198]

As we have seen, these enzyme-operated transformations occur in small energy steps and enable the organism to carry out very considerable chemical changes without any marked rise in temperature. Life is, in the words of Fernel,[6.222] 'a low flameless fire' (p. 625). The reactions occurring in organisms and in the chemical relations between organisms, from mutually beneficent symbiosis to outright ingestion or parasitism, form part of complex, linked, chemical systems. In the fully developed *biosphere*, as it has existed at least for the last 3,000 million years, relatively few organic molecules are permanently set aside; but those that are, such as coal and oil, are of the greatest value to man. Most go round in endless cycles of transformation through plant, animal, and bacterium, back to plant again. The whole biosphere can be considered as one evolving biochemical system. There is no reason to believe that it is the only possible system of the kind in the universe. There may be other biochemical transformation systems on other planets, some less efficient, others more efficient than ours.[6.156; 6.211]

THERMODYNAMICS OF LIVING ORGANISMS

The specific and controlled nature of energy interchanges in living systems, together with the rapid rate of flow of matter through them, go a long way to explain the apparent paradox that they seem to contradict the second law of thermodynamics, which demands that in every closed system the entropy, or mixed-upness, must always increase, or in other words that it becomes less ordered in time (p. 589). Now organisms seem to maintain, over long periods of time, approximately the same degree of order during most of their lives. They actually increase it when they grow and reproduce, and lose it only at death. This was supposed to imply some divinely ordered or purposeful arrangement, but it is now seen as a simple consequence of the fact that a live organism is not a closed but an open system. For such systems, as Prigogine[6.215] has recently shown, entropy does not increase; it merely tends to a fixed value. The second law of thermodynamics is in fact only a special case for closed systems. This knowledge removes any need to consider the thermodynamic aspect of the metabolism and growth of organism as anything specially vital, and does in the twentieth century for organic energy changes what Wöhler did in the nineteenth for organic matter. It does not, however, solve the problem of life; it merely removes a pseudo-problem that had got mixed up with it. It leaves the essential problem, which is that of accounting for the origin and evolution of the ceaselessly changing, but essentially recurrent, pattern of structures and processes that characterize living organisms.

## 11.3 Molecular Biology

The subject now recognized as a separate one of molecular biology arises, on the one hand, out of biochemistry and, on the other, out of the advances in crystal structure analysis. Classical organic chemistry, electron microscopy, and genetics have also contributed to it. The essential feature of molecular biology is that it is concerned with the structure and functions of entities of dimensions below those of the cells studied by biological methods, and above those of the molecules studied by normal chemical methods, in particular, with the molecules of proteins and of nucleic acids. It is now one of the most exciting and rapidly growing branches of biological science. It is very difficult to find a precise date for its start. It might really be said to begin with the early studies of W. T. Astbury (1898–1961) at the beginning of the thirties, on the structure of wool, and it received the first official recognition by the appointment of Astbury as Professor of Biomolecular Structure in 1945.

The whole of biomolecular studies represents an extremely fast and coherent body of research carried out by an informal and somewhat casually organized operation of different fields of work. It can be broken down into the history of the study of the structures of protein molecules, fibrous and crystalline, of nucleic acids and of the viruses which are themselves nucleo-proteins containing both components. These studies led to the great combination which enabled the mutual relations of proteins and nucleic acids to be worked out in principle and have begun to indicate the precise code by which the nucleic acids have determined the structure of the different proteins.

STRUCTURES OF PROTEIN MOLECULES

The early history of protein structural studies began with the more precise elucidation of enzyme action and the preparation of certain enzymes such as urease or pepsin in crystalline form. Already in the twenties it began to appear how essential the role of the proteins was in living organisms. They give at the same time individuality and activity. Compared with most molecules dealt with by organic chemists, the proteins are very complicated. First of all their molecules are big – too big for ordinary chemical methods of measurement but big enough to be amenable to physical measurement, as Svedberg showed when he separated them with the high rotational speeds of the ultracentrifuge, a kind of hundred times speeded-up cream separator.

What was more astonishing was that they could be crystallized, that is millions of protein molecules of the same kind could fit together – 'in rank and file', in Newton's phrase – just as regularly as the simplest atoms in inorganic crystals. This implies that the molecules of proteins of any given kind are substantially identical. The identity need not be an absolute one – down to the last atom or bond – but crystallization does imply that most of the molecules do not differ by more than a few per cent in size and shape.

The existence of protein crystals made it possible to examine protein structure by means of the same X-ray analysis that had previously been applied to organic crystals. This gave exact measurements of the size of protein molecules, which range from those containing 1,000 to those containing millions of atoms – mostly those of carbon, nitrogen, oxygen, and hydrogen. It also gave some clues as to how they were held together.

On the chemical side the essential basis was the work of Emil Fischer (p. 634) which had shown that proteins could consist of chains of amino-acids and provided the most definite clues as to their structures. The first decisive advance on this was the determination by Sanger in 1952, using paper chromatography, of the precise order of the amino-acids in the two chains which make up the molecule of insulin, which led the way to the determination of amino-acid order in many other proteins. This has been so far the greatest triumph of analytical chemistry. However, the chemical analysis did not by itself say much about the most interesting features of the proteins, their capacity for exercising chemical transformations, as do the enzymes, or for forming active physiological molecules responsible for the contraction of muscle, on which all animal movement depends, and for the conduction of nerve messages.

### FIBROUS PROTEINS

Both muscle and nerve are made of fibrous proteins, so are inert parts of animal organisms such as the collagen of cartilages, the keratin of hair, nails, and horn, and the silk of insects and spiders. These hard fibrous proteins may be considered in some sense biological by-products, excreta retained for structural purposes. Fibrous cellulose plays the same role in plants and chitin in the hard skins of insects. It is just because they are solid, strong, and stable that the fibrous proteins have proved to be of value to man from primitive times, and have become the basis of the great wool, silk, and leather industries.

For the same reason they were the first proteins to be analysed by means of X-rays. The work of Mark and Astbury showed them to be chains of amino-acids which seemed to be folded or coiled in elastic proteins like wool, and were straight in rigid proteins like silk. This has done much to give a scientific basis from which to modify the old techniques and to provide means for the creation of new textile fibres. Already a new range of secondary fibrous proteins has been produced from natural globular proteins, such as ardil from the edestin of ground nuts; and truly synthetic proteins, like polybenzoyl glutamate, can now be made in the fibrous form, and threaten to rival the completely artificial polyamides of nylon.

### STRUCTURE AND GENESIS OF GLOBULAR PROTEINS

It is, however, a far cry from the artificial production of fibrous proteins from amino-acids to the actual construction of active, so-called globular, protein molecules such as those found in crystalline proteins.

The fibrous and globular proteins, however, were not so far apart as their properties seemed to indicate because there are many cases where one can be converted into the other: in insulin, for instance, where a fibrous form can be prepared, and in the actin of muscle from which a globular form of this normally fibrous protein can be prepared. The clue to this transformation was given by Pauling's theory of a helical arrangement of amino-acids stabilized by internal hydrogen bonds, forming what we know now as a *secondary* structure of protein molecules, both fibrous and globular.

The concept of helical structure without complete crystalline regularity proved to be an extremely fertile idea, not only for the structure of proteins but also for those of viruses and nucleic acids. It led to a new burst of analytical activity which resulted in the triumphant successes in the complete molecular analyses of all these forms.

MYOGLOBIN AND HAEMOGLOBIN

The attack on the crystalline proteins was first brought to a successful conclusion by Kendrew and Perutz in using classical crystallographic methods to fix the repeat patterns by the introduction of heavy marker atoms to fix the phases of the X-ray reflections from which atomic positions can· be calculated (pp. 738 f.). The first rough analysis of a protein by Kendrew showed that in sperm whale myoglobin (an odd commentary on Moby Dick) the molecule consisted of connected Pauling helices folded around the central iron-containing porphyrin, the 'business' part of the molecule for holding oxygen. This is a so-called *tertiary* structure which is a specific determinant of a chemical function of a protein and is beyond the reach of normal chemical methods of analysis.

Later studies showed, in advance of the chemical sequence determination, what the arrangements were of the atoms themselves. In the more complicated structure of haemoglobin, Perutz showed that there are two pairs of almost identical sub-molecules.

The importance of precise structure was revealed when it was shown that the slight difference produced in haemoglobin by the substitution of one amino-acid caused the particular disease 'sickle-celled haemoglobin', the first genetically-controlled *molecular disease* to be recognized.

The study of protein structures is still only at its outset. It may be reasonably inferred that the *tertiary* structure, or way in which a necklace of amino-acids forming an enzyme molecule is curled up, has a great deal to do with the attachment of the substrate molecule on

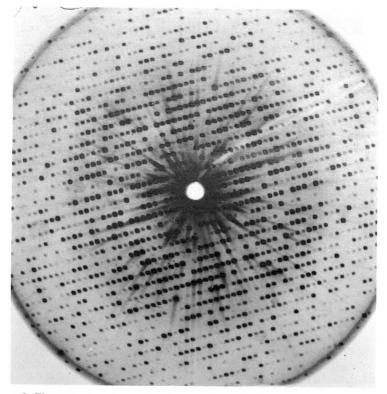

278. The molecules of organic substances found in living creatures are extra-ordinarily complex. The use of X-ray crystallographic techniques of J. C. Kendrew and M. F. Perutz has helped elucidate the structure of myoglobin and haemoglobin, both of which are protein substances (see page 907). This crystallograph shows myoglobin. And its complexity compared with a simple inorganic substance with smaller molecules can be seen by comparison with illustration 6 above.

which enzyme action has to be operated: it is a kind of jig. But so far no structure of an enzyme protein has been fully worked out. Nor is there yet a classification of proteins based on their structure.

It would also appear, however, that all the complexities in a globular protein are determined by the sequence of amino-acids, particularly the bridges which are made between very different parts of the molecule by means of sulphur–sulphur links. The comparative study of protein structures has hardly begun. The presence in cells of a number of these different forms of proteins may ultimately provide clues to the history of the development of protein and nucleic acid synthesis (pp. 912 ff.).

## NUCLEIC ACIDS

The unravelling of the structure of nucleic acids started more slowly than that of the proteins, largely because nucleic acids are not found, as are some fibrous proteins, in a relatively pure state. Their name indicates them as contents of the nuclei of the cells. They were found most abundantly first in rapidly dividing yeast and then in the thymus gland prominent in children in their rapidly growing stage. The association with growth and protein formation was emphasized in the work of Casperssen in the thirties. The fact that they absorbed ultra-violet light and took up certain dyes marked their presence in large quantities in the chromosomes, already known to be associated with genetic changes and reproduction.

Chemically they were shown to be a complex of purine and pyrimidine nitrogenous bases, linked with a specific sugar; either ribose in so-called ribose nucleic acid (RNA) or yeast nucleic acid, the closely related deoxyribose in deoxyribonucleic acid (DNA), thymus nucleic acid. The sugars in turn were linked by phosphate groups and the nucleic acids seemed to be the main phosphate carriers in all living organisms.

Their serious structural study was begun by Astbury in 1932 after they had been isolated and found to be dissolved in a sticky fluid which could be drawn into threads indicating a fibrous polymer struoture. Astbury showed that the four nucleosides, the purines, adenine and guanine, and the pyrimidines, cytosine and thyanine (uridine in RNA) piled up like coins at right-angles to the axis of the thread. Furberg showed that the ring of the sugar molecules was set at right-angles so that it could be joined up through the sugars by the phosphates to form a polymer.

When the chemical analysis of Chargaff showed the number of purines and pyrimidines were exactly balanced, Crick and Watson put their famous hypothesis of the arrangement being not a *single* helix but a *double* one, the purine of one chain being linked with the pyrimidine in the one twined with it. This was subsequently verified by X-ray analysis by Wilkins and Franklin (1920–58). Though even nucleic acids contain all four nucleosides, their precise *order* is what is characteristic for each specific nucleic acid and is quasi-automatically transmitted when a new but identical molecule of nucleic acid is laid down on the

279. The complexity of organic molecules lies not only in their size but also in the way in which the component atoms are arranged. A model of the DNA molecule with its double phosphate-sugar chains of atoms is here being demonstrated by the geneticist K. Mather. B.B.C. photograph.

280. A model of the DNA molecule showing the double phosphate-sugar chains surrounding base pairs.

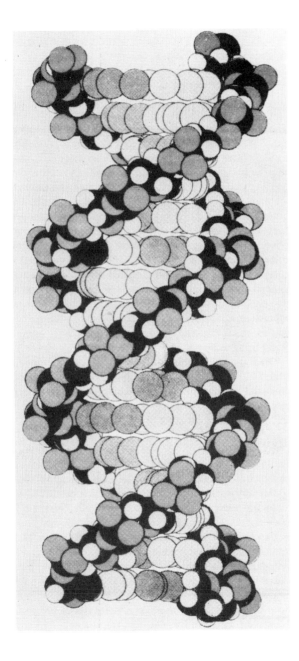

coil of the old one. The picture of this structure of the nucleic acid molecules contains all that is necessary, in principle, to permit an information-carrying and transmitting tape to be built into the very inner core of every cell or virus particle.

## MOLECULAR REPRODUCTION

The implications of this purely structural discovery were enormous; it was the greatest single discovery in biology. It enabled the structure to be linked immediately with the whole body of knowledge which had been accumulated for decades in genetic and chromosome studies.

The cytologists had long suspected that the essence of reproduction, which apparently to the microscope starts in the chromosomes, consisted in a duplication of sub-molecular structure and by simple logic they showed that this could not be a *bodily* duplication or even one of plain reproduction such as that of printing, because that always inverts right and left, but it could only be a linear reproduction in which a line is repeated point by point along its length. Now a helix is essentially a twisted-up line and the accurate pairing of purines and pyrimidines gave the possibility of having a self-reproducing line which would enable precise reproduction of the nucleic acid molecule to take place. That this is effectively the mechanism was proved later by Meselson and Stahl using isotopes.

Once the key to molecular reproduction had been found, the way it was to be used in protein production and through that in the whole chemical mechanisms of life did not have to wait long. It came essentially through study of the viruses which represent life in its simplest form, but not in its most primitive.

## VIRUS STRUCTURES

Among the viruses, we find a whole range, from the relatively large and complex animal viruses that cause such diseases as measles and smallpox to the very small viruses that cause the innumerable diseases of plants, and some such as poliomyelitis in animals. There are viruses of even the bacteria themselves, the bacteriophages, the last link we may imagine in the chain of the 'larger fleas having smaller fleas on their backs to bite 'em'. In their action in causing disease, which can be transplanted from one organism to another and may even lead to epidemics, the viruses differ in no essential way from the bacteria; in fact they were distinguished from them only by their ability to pass through filters used to hold up bacteria and their invisibility under the ordinary microscope. Now that we have the electron microscope, viruses can be seen and much of their gross structure distinguished. Even earlier Bernal and Fankuchen had shown by X-rays that the first virus to be isolated by Stanley and by Bawden and Pirie in 1934, that of tobacco mosaic virus (TMV), was a rod-shaped body with an inner regularity. In 1954 this was shown by Watson, Wilkins, and Franklin to be a tube constructed of protein molecules into which was woven a thread of nucleic

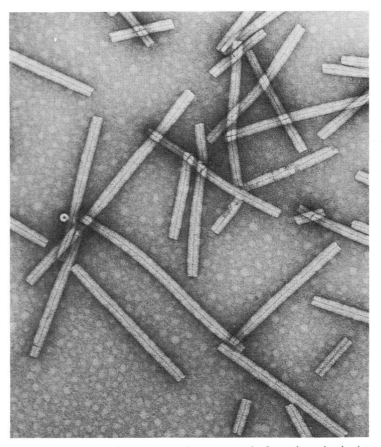

281. The tobacco mosaic virus is a tube constructed of protein molecules into which is woven a thread of nucleic acid. The magnification on this photograph is 220,000 times.

acid. In the apparently spherical viruses the protein molecules formed a regular polyhedral cage for the nucleic acid. Yet viruses differ in one essential respect from bacteria: as yet they have not cultured on artificial media. Suspensions of them can be kept without detectable metabolism for years yet remain fully virulent. Connected with this is their apparent high resistance to antibiotics. The protein covering of the virus particle was found to be a mere protection, the generative part was a nucleic acid which, once it had entered a cell, reduplicated, by the process already referred to, at the expense of the whole of the metabolism of the cell. This was turned over, under the orders of the virus, to make more virus, and later, more virus proteins usually to the extent of completely destroying the cell and leaving behind a crop of new virus particles ready to infect other cells. It was seen that the virus had a capacity for ordering reproduction of a specific protein that fits its nucleic acid and *only* its own nucleic acid. It was capable of taking charge of the directive metabolism of the cell and turning it to its own use.

By the very fact that they could reproduce, it was evident that viruses were organisms; it was equally evident that they were not complete organisms, they were molecular parasites of organisms, a kind of cuckoo. The hosts might be various, but the virus always retained its identity. Yet this crude cell destruction by the virus was outdone by some viruses that were able to enter and join themselves up with the genetic mechanism of the cell itself and so be transmitted with the division of the cell from cell to cell. Most of the time this has no other effect, but from time to time the virus nucleic acid breaks out and reproduces the virus from an apparently healthy plant. This process may well be related to the genesis of at least certain cancers (pp. 934 ff.). The problem immediately posed itself 'what was the way in which the virus nucleic acid controlled the protein-synthesis mechanism of its victim?'

DECODING NUCLEIC ACID

The DNA-containing viruses have proved to be the most useful tool for finding the nature of the code by which the order of the four nucleotides can provide the information needed to build a specific linear protein chain containing the twenty-four amino-acids in the appropriate order. This can be done because it is possible to get two virus DNA molecules to couple inside the bacterial cell. If these carry different mutants, or what we might call misprints in the nucleotide chain, the nature will be revealed by the proportions of the different virus offsprings they produce. In this way it was possible for Crick and Brenner to show that the nucleic acid code operated by 'words' of three nucleo-tides to correspond with one particular amino-acid and that some of the arrangements could be determined analytically. This work could be complemented by synthetic studies. Already it had been shown by Ochoa that, by a fairly simple polymerization reaction, the nucleotides could be built up into nucleic acids, either of one or two or of several nucleotides and the particular one built up could be determined by introducing a piece of complete nucleic acid as a kind of template which could then be reproduced indefinitely. In other words, reproduc-tion of recent forms of life had been achieved in the laboratory. The crucial experiment was made by Nirenberg and Matthaei in 1961 in which the introduction of the artificial nucleic acid, poly-uridine, led to the production of the artificial protein poly-phenyl-alanine. The decoding process is now going on vigorously in many laboratories although it is evident that the mere scale of the work will require the means of transferring the whole process, experiments and all, to com-puters before we can get really practically useful information.

## NUCLEIC ACIDS AND PROTEIN SYNTHESIS

Although it was already evident by 1953 that nucleic acids contained the code for protein synthesis, how they operated required further investigation and revealed unsuspected but soluble complexities. Their studies brought to light first a kind of local hierarchical system in which the DNA lay in the nucleus and the actual synthesis of proteins took place in the exterior cytoplasm located in small bodies, the so-called ribosomes, containing RNA. The mechanism revealed by the work of Volkin and Astrachan showed that the first stage in making a new protein was the reproduction on the DNA in the nucleus of a special form of RNA, so-called *messenger* RNA, containing the necessary information for constructing a particular protein. These messenger RNA molecules travelled out into the cell cytoplasm. Already assembled there was a series of amino-acids each labelled with a specific *soluble* RNA component, that is with a short length of nucleic acids. The soluble RNA molecules then attached themselves in the appropriate place on the messenger RNA molecules thus ensuring the correct order of their accompanying amino-acids to make the specific protein.

Never has Oscar Wilde's dictum of nature copying art so well been exemplified, for this is practically a reproduction of a modern auto-matic assembly line on a molecular scale, the whole process being tape-controlled on a self-reproducing tape. When it is considered that this has to be done in every cell to at least two thousand different proteins and without a single mistake – for a single mistake means a mutation which is normally lethal – you get some idea of the real complexity of life which was beyond the imagination of anybody, even in science, as recently as twenty years ago. The work of establishing even sketchily and provisionally this elaborate picture represents one of the greatest triumphs of human thought. It is one which characteristically required an enormous degree of intercommunication and co-operation of scientists in many fields and which is only at the beginning of its course. Further triumphs and elucidations can be confidently expected within the next years or even months.

The mechanism of protein formation in the cell seems to be limited to the formation of the primary structures, that is, the order of amino-acids along the chain. The code merely ensures that the expression of order in the four letters of the nucleic acid series is reproduced in the order of the twenty-four letters of the amino-acid series. Subsequent formation of discrete protein molecules is implied, apparently, by the order of the primary structure, that is, the secondary and tertiary structures follow from the primary structure. To a certain extent the secondary or tertiary structure can be destroyed and reconstituted by purely physical methods. This means, effectively, that a linear code can be transformed into a three-dimensional one.

Further, all the more complicated structures in the cell seem to be laid down in terms of protein units put together; fibres like collagen and muscle; intracellular membranes such as nuclear membrane and those of mitochondria have a protein element. Perhaps most important of all the organelles in the cell, such as centrosomes and their associated flagelli, seem to be all made on a predetermined structure, predetermined by the shape of the identical protein molecules which constitute them and, apparently, from which they build themselves up spontaneously.

One clue to this autosynthetic process had been provided by knowledge of the way in which the protein envelope of viruses is built up. Most viruses seem to make only one kind of protein and this fits together into a shell, either a helix or a little polyhedron, which is the typical envelope. But these envelopes can be disintegrated by purely chemical means, the nucleic acid or 'business' part of the virus destroyed and then, by chemicals again, the virus structure can be reconstituted, apparently identically, into what we call top or empty virus which is also often formed naturally.

This is an indication of the basic importance of the protein-building mechanism in producing absolutely identical molecules which can then fit together in a limited number of highly specific ways, thus making the complicated structures. The very complicated structures such as the $T_2$ bacteriophage seem to have five different kinds of protein molecules, each controlled by a different gene in the DNA of the virus itself.

The building of the virus itself, with its complexity, reveals that it is sufficient to make a set of different series of amino-acids: they would first coil themselves up into protein molecules and then arrange the protein molecules mutually without any other interference of a vital character. This enormously simplifies the problem of structure formation in biology, though we may be only at the beginning of a long story. The same kind of mechanism may well explain – although this remains to be proved – the formation of the complex organelles in cells, each controlled by the existence of a set of protein molecules themselves genetically determined.

Nor is the inquiry an idle one, for already it has its fruits in medicine. These elaborate processes, once they are understood, can be controlled. Nucleic acid formation can indeed be cheated by introducing nucleotides of great similarity to the natural ones but sufficiently different, so to speak, to block the works. This is the basis of certain new anti-viral drugs such as thiouracil and may in the end be the securest route to the conquest of cancer.

It would seem at first that we are approaching the ideal of the alchemist of actually making life, the homunculus in the adept's globe. But, in fact, the finding of the code only shows what an extraordinarily complicated process this is going to be. The more we know about life the more difficult it would appear to make it to any *good* purpose. That it could be made to a *bad* purpose is only too evident. We may even expect in the present world that the first live particle we are likely to produce artificially would be a virus of a hitherto unknown disease, an eminently suitable occupation for scientific workers in biological warfare.

THE SIGNIFICANCE OF MOLECULAR BIOLOGY

All these discoveries are so recent that they are, despite modern publicity, not appreciated even in scientific and still less in other intellectual circles. The idea that the identical nature, not just of man as a species but of any individual, is determined in fullest detail on a small molecular code line which is only about a quarter of an inch long, a molecule that is a few millionths of an inch wide, exceeds the dream of the most absolute fatalist. It is at any rate an enormous step towards what has been called the 'secret of life', and is now appearing as just the first unveiling of secrets inside secrets.

We can now turn round and define life in these molecular terms; extending Engels we could say 'life is the mode of motion of the protein-nucleic-acid combinations'. Even if clearly not true for life at all times, this is certainly true for life on this earth as we know it and find it now, which is one biochemically interconnected unity every element of which, down to the smallest virus, operates its synthesis by this particular molecular mechanism. This will be enough for the biologists, molecular and gross, to deal with for a long time to come, but it still leaves the question of the origin of life unanswered. However, it would be most unscientific to imagine that such intricate and beautiful mechanisms came suddenly into existence. To suppose so is just the same kind of mental laziness that led primitive men, with far more excuse, to compose creation myths. Any alternative to these is found to be difficult, but approaches to a theory of the origin of life have already been made (pp. 984 ff.).

## 11.4 Microbiology

The basic chemical nature of life can best be seen when it is not complicated by the elaborations of form and behaviour. Biochemistry in the twentieth century is at last beginning to unravel the secrets of the life of the smallest of organisms, of bacteria, yeasts, and moulds, and of the simplest of animals – the single-celled protozoa. The simplicity is one of form and structure alone; biochemically, as we shall see, they are at least as complex, if not more so, than the higher organisms. Strong incentive and support for their study have come both from medicine, in the treatment of the diseases they cause, and from industry, because of the chemicals and drugs they produce, including the most important one, the universal drug, alcohol. Their contribution to agriculture is now also beginning to be studied, for on them depends in large measure the fertility of the soil.

### CHEMICAL VERSATILITY AND ADAPTABILITY
### OF SIMPLE ORGANISMS

We are only now beginning to glimpse at the possibilities of microbiology when it is approached by chemical methods. Much can be learned about the normal and abnormal life-processes of these minute organisms by growing them in solutions containing a variety of substances. The effects of these on their growth can be studied, and information can be obtained as to the transformations they undergo in the organism by examining the products excreted into the medium. What emerges from these studies is that the morphologically simplest of organisms are of the highest chemical complexity. Indeed they are able to carry out any process that is achievable by higher organisms, and often many more. They seem to be like little chemical factories in which molecules are passed along the line from one enzyme to another to be incorporated into the organism as growth, to have energy extracted from them, and finally to be excreted as unusable remnants. Different organisms specialize on different processes, but, somewhat unexpectedly, the specialization appears not to be by any means rigid. The metabolism of simple organisms seems remarkably adaptable.[6.192] If one food molecule is not present they soon make use of another, and change many of their chemical processes in order to do so. This

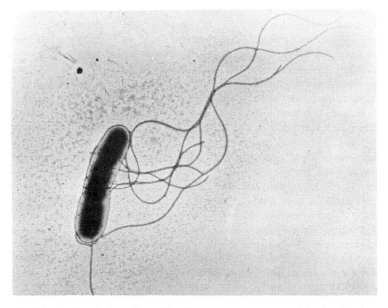

282. The bacillus *Salmonella typhimurium*, a common cause of food poisoning, showing its thread-like flagella, the appendages responsible for its mobility. An electron-micrograph with a magnification of 27,000 times.

variability is often most annoying to us because it also works for anti-bacterial poisons, and many strains have now got used to sulpha drugs, and some even to penicillin. It is formally a kind of chemical learning, and once we have mastered its mechanism we shall be able to teach these organisms to make what we want. It shows a toughness and flexibility in primitive organisms that have enabled them to survive and evolve in the processes.

AUTOTROPIC BACTERIA

The apparently simple but probably degenerate viruses, already described, may be taken as the extreme of organisms or organelles absolutely dependent on a cellular metabolic environment. Moulds and protozoa are relatively complex organisms with internal structures visible under the microscope. Even the simpler bacteria have characteristic forms and are beginning to show internal structure in the electron microscope. All of them have quite an elaborate metabolism if placed in a suitable medium. As we have seen the smaller and simpler organisms, the viruses, lack even that.

At the other extreme in the scale of chemical behaviour, representing absolute independence as against complete parasitic dependence, are the autotropic bacteria, such as those living in the soil and in hot springs, which can satisfy the whole of their needs with simple salts such as nitrates and sulphates. Some do not even require oxygen to live on, but make up for it by oxidizing and reducing iron and sulphur compounds. They are of considerable economic importance, as most of the sulphur deposits are made by them. Their extreme self-sufficiency shows that they must be much closer than the viruses to really primitive organisms. Nevertheless they cannot be really primitive, for they are completely sophisticated in their internal chemical equipment, having not only all the enzymes that other organisms have, but a few more necessary to deal with the simple substances on which they feed.

It would appear that primitive bacteria have developed into other organisms having less, rather than more, chemical adequacy if taken separately. An autotropic bacterium can live in an entirely inorganic environment. All animals and many plants have lost some of these mechanisms, and are dependent on the environment for already organically prepared food or auxiliary food substances such as vitamins.[6.196] The more primitive of these organisms live simply on the products of decay or secretions of other organisms, which they ingest through their cell membranes. Others, slightly more advanced, have found a way of moving by means of mobile threads called cilia or flagella into regions where there is more food. Others, still single-celled, such as amoeba, have taken the next decisive step in actually ingesting pieces of food – either living or dead matter; that is, in living effectively parasitically on other organisms. Now this tendency has a twofold effect. In the first place the mere availability of food derived from the bodies of other organisms, which contains many essential substances already formed, removes the necessity for many biochemical processes which the more primitive organisms require. They become therefore simpler chemically, but only by becoming correspondingly more complex organizationally and functionally. They must be able to react to food situations and not merely vegetate; they must be able to move to where there is more food and have some way of catching it.

## THE IMPORTANCE OF SIZE

For this, size is an important factor. Small single-celled animals can manage quite well in their immediate neighbourhood – they need no organs of movement to get around. On the other hand, if they grow bigger, the effort of getting around, and even more the business of taking in the food for the whole organism at one mouth, becomes very difficult. There are two solutions, in principle rather different. One is for the organism to stay still and sweep the food past it; in a primitive way this is done by the sponges; in a more complicated way by oysters and barnacles. The other is to go after it; that is the way of the fishes, the reptiles, and finally ourselves. We have taken the still further step of actually persuading other organisms to produce our food through the processes of agriculture. The general trend of evolution is away from the purely chemical existence of minute units to the use of increasing organization, co-ordination, and rationality.

## THE UTILIZATION OF BIOCHEMICAL PROCESSES

Even when armed with greater knowledge of the origin of life (pp. 984 ff.) it is still unlikely that we shall be able to create life artificially. What is much more likely, and what may even be achieved in a few years, is the effective carrying out of many of the functions of life for our own benefit by purely artificial means, particularly that essential function the photosynthesis of organic materials. If we could use the sunlight which strikes the earth today and turn it directly into human food without the intervention of plants, a major underlying problem of world economics would be solved at a stroke, and the possibility of an unlimited expansion of the human race would be assured. Here again we can see the link between the acquisition of knowledge and that of

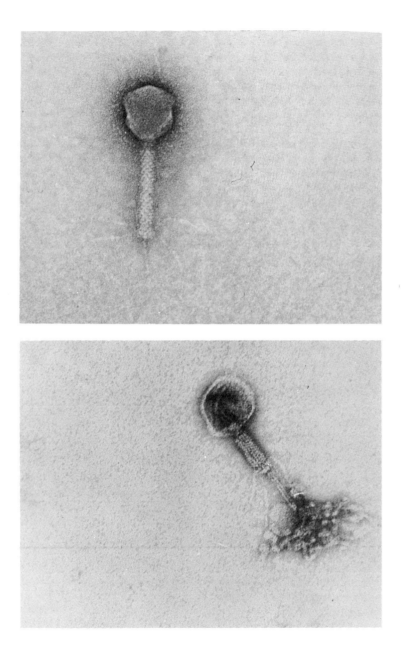

283 a, b. Two views of a bacteriophage particle. This is essentially a virus that attacks bacteria (see page ooo). Under very high magnifications with the electron microscope (magnifications of some 370,000 times) the particle is observed (illustration 283a) to have a head and tail; the head contains a nucleic acid destructive to a bacterium. The tail possesses fibres by which the bacteriophage attaches itself to a bacterium. Illustration 283b shows the same type of bacteriophage after it has attached itself to a bacterial cell wall and injected nucleic acid down through the tail. The head is now empty.

power. Before we can hope to reproduce any of the characteristics of living organisms we must first understand how the living organism manages itself; and that will mean a great deal of research, most of which will be directed not at solving that problem but simply at finding out the relations which may later be used for solving it (pp. 983 ff).

## 11.5 Biochemistry in Medicine

Now, as has already been pointed out (pp. 876 f.), the original impetus for biochemical research came from medicine. As physiological chemistry it became important in the twentieth century, largely because it marked the second stage of the great medical revolution heralded by Pasteur's work in the nineteenth. The early bacteriologists were content to proceed in a purely biological way, using for remedies vaccines or antisera prepared from the bacteria themselves. Later the desire to obtain more certain results led to a deeper study of the chemical mechanism of these treatments. This blended with another stream of research which emanated from the study of deficiency diseases and of disorders of metabolism, which were also found to have a chemical basis. Biochemistry was the common link that bound them all together.

The more scientifically disease is studied, the more does it appear that it is associated with abnormal biochemical behaviour of cells and tissue fluids, with an interference with the balanced equilibria of molecular transformations that we call living. The interference may be gross, as when an injury or a swelling breaks some vital connexion and cuts off supplies completely, as in gangrene and pneumonia, or it may be insidious, as in the degenerative changes producing diabetes. The body, or any part of it, is said to be diseased if it lacks some chemical it needs or acquires some that interfere with its working.

Apart from purely mental afflictions, all diseases are in the last resort due to starvation or poisoning. They fall into groups according to how the poison enters or why the needed substances are absent. These four groups are not exclusive, because one may lead to another and, unfortunately, it is possible to have all of them together. They are: (1) the infectious or parasitic diseases; (2) the diseases of deficiency, external and internal; (3) the diseases of faulty tissue-growth or cancers, which may well, when we know more of them, prove to fall under groups (1) or (2); and finally (4) the diseases in which mental disturbances of social origin may upset the chemical balance of the body. In the prevention and cure of diseases in all these groups, but especially in the first two, there have been spectacular advances in this century, and most of them in the last two decades.

This classification of disease was made provisionally here to bring out twentieth-century advances in the understanding and control of disease through the use of biochemistry. It is not intended, however, to give the impression that disease is merely the upset of a chemical balance in the body, to be put right by a specific chemotherapeutic substance, or in plain language by a new medicine out of a bottle. This advance is nevertheless an important one. It has enormously helped the battle against disease by providing the doctor with new tactical weapons, but it is no substitute for the general strategy of a long-term campaign for health. For this involves the whole human being and his economic and social environment. Good food, clean work, companionship, and an active and reasonable faith in the future are the basic essentials. Without them, all the triumphs of biochemical science are mere palliatives; with them, they provide more and more successfully against contingencies of external infection or internal deficiency.

### ANTIBIOTICS

In dealing with infectious diseases where the cell-poisons are produced by foreign organisms living in the body, twentieth-century medicine, while maintaining and refining all the methods of Pasteur, has moved one stage further. It is still as necessary as ever to prevent germs and parasites entering the body, but now they can be dealt with increasingly successfully even after they have done so. The attempt to do this has given an enormous stimulation to the study of the direct action of specific chemicals on micro-organisms and on their hosts, particularly man. Although the original motive has been that of conquering disease, one very important, and certainly far more generously financed, motive

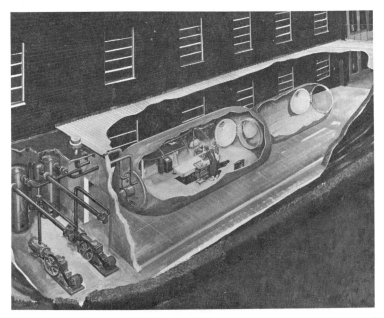

284. Even with the use of antibiotics, it is still necessary to prevent the ingress of germs to the body during surgery, and designs are now being made for hyperbaric operating theatres which are, in effect, pressurized chambers. Such theatres appear to have added advantages from the medical point of view both during surgery and in post-operative treatment and the maintenance of hyperbaric conditions may be advantageous enough to lead to the establishment of pressurized wards. A design for an hyperbaric operating theatre by the Hercules Powder Company of Delaware.

for these studies has been that of causing diseases, either by poison gas or now by radio-active poisons and mass bacterial attack.

Ever since Pasteur discovered bacteria there was always the hope that some chemical could be found that would kill the bacteria inside the patient without also killing the patient. Where the infective organisms were of a kind very susceptible to chemicals, such as the trypanosomes of sleeping sickness or the spirochaetes of syphilis, there was some hope that simple inorganic compounds, especially those of the heavy metals, might have a good effect. This had already been found to be so in the nineteenth century (p. 634), but the common run of diseases caused by bacteria had proved much more recalcitrant.

The first success was arrived at by trying to see whether chemicals that would dye bacteria for recognition purposes could also be used to track them down in the body and kill them. This was the origin of the first group of chemotherapeutic substances, the sulphonamides, first produced in 1932 by Domagk (1895–1964).

## PENICILLIN

It was not long after that the epoch-making discovery of penicillin was made. This discovery is an extremely good example of the strength and weaknesses of scientific organization in the twentieth century. Fleming (1881–1955) in 1928 noticed that some of his bacterial cultures were being eaten away at various spots, and was a good enough observer to note that this was due to the appearance of a mould on his slides which seemed to be giving off some substance which killed the bacteria. The mould was wrongly identified by the mycologists, and for about ten years nobody thought it was worth following up. This does not mean that no one would have been interested in this observation if they had known of it; on the contrary, there were very many people looking for any non-toxic substance that would destroy bacteria. What was lacking was an organization to search for and develop any promising openings. It was not till ten years later, when Florey (1898–1968) and Chain, stimulated by the success of the sulphonamides, started a systematic search for natural antibiotics, that Fleming's observation was put to use. The great efficacy of extracts from *Penicillium notatum* led immediately to a concentrated chemical attack to separate the active principle, and to show that it was poisonous only to bacteria and not to their hosts. The experiments on animals were so promising that efforts were made to prepare enough of the drug for the treatment of human beings. This was necessarily something of a gamble, because the value of the drug could be proved only if enough could be got to follow through serious cases to complete recovery, and then to treat enough additional cases to show that it was not just luck.

By the time the clinical value of the drug was proved the War came, and the subsequent stages of its purification and large-scale preparation were rushed through at a rate that could never have been achieved in peace-time. It was a concentrated effort in the fields of chemistry, biology, and medicine, on a scale of brain-power comparable to that devoted to the atom bomb. It was a hurried job, employing probably far more scientific workers than were strictly necessary, but it was done. Had it been left to go the slower way many man-hours would have been saved, but thousands of people would have died. It is also by no means certain that, but for the war, penicillin would ever have been developed at all. It did not seem particularly promising at first, and it would have been difficult to raise the funds to push it to the point of proved value. After penicillin had been made, three further tasks remained to be done: to find out what it was; how to synthesize it; and how it worked in destroying bacteria. The first was accomplished in 1944: the discovery of the detailed formula of penicillin was largely due to the use of X-ray technique;[6.164] the second has so far baffled the chemists; on the third some advance has been made. It is by far the most important task of all, to find the mode of attack of a chemical molecule on a bacterium, because once that is found it should be possible to design a molecule that works as well or better, and is far easier and cheaper to make. There is some evidence now that the efficacy is due to the molecule of the antibiotic being very but not quite like that of the normal food of the bacterium, so that it is taken in and jams the works.

CHANCE AND PLAN IN SCIENTIFIC ADVANCE

The discovery of penicillin is often used to prove that important discoveries come by chance. The answer is that the particular combination that does the trick does come by chance, but that chance is multiplied by providing opportunities for discovery in the first place, and for development by interested people in the second (pp. 607 f.). Once penicillin was discovered it was relatively easy to search through Nature for other substances which might have the same or better effects, and a whole new field of antibiotics was opened – streptomycin, aureomycin, chloromycetin, etc., etc. Even now, however, the hunt for antibiotics resembles a gold rush rather than a properly conducted scientific prospecting operation.[6.172] The scientists and the pharmaceutical firms backing them are so keen on getting out a new antibiotic that they sacrifice the possibility of fundamental discoveries as to the genesis and mode of action of antibiotics in a feverish search among a large group of organisms for anything that will work. It is characteristic of the attitude of monopoly capitalism towards discovery that, whereas all the work on the production of penicillin in the first place was carried out by British doctors and research workers, who published their results freely, the actual manufacture of penicillin is covered by US patents, and thus every unit of penicillin used in the country of its origin has to pay royalties to American chemical firms.

THE ORIGIN OF DEFICIENCY DISEASES

The main outlines of the problems of the second group of diseases have already been given in the discussion on vitamins and hormones, the discovery of which was one of the major achievements of twentieth-century biology (pp. 896 ff.). From these studies a more general picture of the chemical behaviour and control of organisms is beginning to emerge. The higher animals and plants have evolved from simple forms that were probably as generally competent chemically as bacteria are today. They could make all the complicated substances they wanted from simple inorganic molecules. When the organisms got more complicated some of their cells ceased producing many specific substances – mainly co-enzymes like vitamin $B_6$, or nicotinic acid – as well as some more complicated hormones such as insulin. This did not matter, as they had also evolved circulation systems, so that a few cells specializing in their manufacture could make enough for the whole organisms. Animals and some plants like the fungi went further: they took in organic matter wholesale as food, vitamins, and all, so that they no longer needed to make them themselves. No harm resulted as long as the food supply was adequate and nothing went wrong with the groups of specialized cells or glands. But if either happened, the other cells, which had lost the chemical elasticity of simple organisms, were increasingly damaged, and ultimately the weakest of them would give way and the whole animal would die.

CHRONIC DISEASE AS METABOLIC DEFICIENCY

After the successes early in the century of the understanding and cure of such external deficiency diseases as scurvy (vitamin C) and beriberi (vitamin B), and internal deficiency diseases such as goitre (thyroxin) and diabetes (insulin), it began to be apparent that a very large number of chronic diseases were deficiency diseases, though in some cases the deficiency might be the effect of earlier infection. This was a challenge to track down the missing substance that could counteract them. The latest successes have been in pernicious anaemia (vitamin $B_{12}$) and arthritis (cortisone and ACTH). We still need research to find out whether the general tissue and arterial hardening or the abnormal fat deposits that lead to cerebral haemorrhage and heart disease are due to the lack of some hormones or the presence of some toxic substances in food.[6.154]

Success in this field may be as important in the twentieth century as that in the case of acute infectious diseases in the nineteenth, particularly as the diseases are predominantly those of later life. In modern industrial populations a larger proportion than ever before are elderly, and, if age could be freed from the disabilities and premature deaths due to chronic disease, human happiness and effectiveness would be enormously increased. In actual life diseases do not fall so neatly into categories. Infections produce deficiencies, deficiencies make the subject more liable to infection. Both are affected by housing and working conditions and by psychological and social influences. The problems of health will always remain much greater than anything medicine or biochemistry alone can do to solve them. Yet without biochemistry no serious solution is possible.

A BIOCHEMICAL INDUSTRY

The successes of biochemistry in medicine and agriculture have now, by the middle of the century, given rise to a new and important industry, that of fine chemicals (p. 828). And what we have seen is only a beginning. Enormously more could be done, and done quickly, by

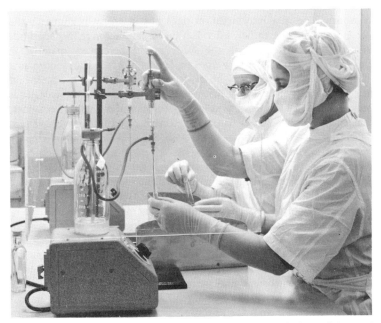

285. Ampoule filling of antibiotics requires both specially hygienic conditions and the delivery of precisely regulated quantities.

devoting far greater effort to chemotherapy research and by building on it an industry which, more than any other, ought to be under public ownership, for it holds the health and lives of people in its hands. Such an industry would not operate merely by conventional chemical means: it would necessarily tend to become more and more microbiological, linking on the one hand with the traditional brewing and baking industries and on the other with agriculture.

## 11.6 Cytology and Embryology

The microscopic study of the structures of cells long anticipated those of cell biochemistry and still longer that of molecular biology. Nevertheless, it is on the lines of molecular biology that we can now interpret and extend the observations of cells as a whole, either isolated in monocellular animals or tissue culture or together in a living higher organism. The new interpretation does not in any way diminish the value of the old but, rather, extends it through new and refined methods of cytology, particularly in the study of the hereditary material in chromosomes, so that it now appears less as a mystery and more as a link between natural history observations, on the naked eye scale and the minuteness of chemical investigations, on the atomic scale. It is only in the last fifty years, and largely through the study of enzymes, that chemistry has begun to be an effective way of approaching biological problems. The earlier contacts between biology and chemistry were invaluable aids to the progress of chemistry but made little contribution to biology. The arguments used in Darwin's *Origin of Species* did not depend on any chemical knowledge. The methods of observation and dissection have also, in the twentieth century, made enormous advances, pushing forward step by step the limits of microscopic vision. First by observation alone, and later by observation combined with experiment, the inner structure of the cells was gradually elucidated. The nucleus with its chromosomes, and the cytoplastic inclusions such as the mitochondria and plastids, were all studied, both in the resting centrosome and more fully in the dividing cell, though they were on the very limit of what could be seen on the optical microscope. Interest in them was enormously increased in 1910 when Morgan showed that the chromosomes of the cell were closely connected with the inheritance of specific characters already forecast in Mendel's theory of heredity (p. 951).

The development of physics had in the meantime brought into existence a number of new instruments. The old optical microscope had remained relatively static for the sixty years before 1940. Now a new and far more powerful microscope was available in the electron microscope (pp. 785 f.). This has been supplemented by some new modifications of the ordinary microscope, which were actually stimulated by the competition of the electronic instrument. The most important of these were the phase and interference microscopes, which enabled cells to be studied alive when previously they had to be killed and stained; and next came the new ultra-violet and infra-red reflecting microscopes, which brought out detail not otherwise visible and could also be used to study the chemical composition of cell structures.

These show the cell to be an enormously complicated but, at the same time, ordered structure. It now appears that it consists of an assembly of different types of even smaller distinct parts, or organelles, whose structure is now known approximately down to molecular dimensions. Some contain nucleic acid, as do the chromosomes of the nucleus and the microsomes or ribosomes of the cytoplasm, whose role in reproduction and protein synthesis has now been elucidated. Others, such as mitochondria, are concerned with enzymic-metabolic activities. With the mitochondria we must now include the lysosomes that do for catabolism what the mitochondria do for anabolism or building up the general respiration of the cell. The lysosomes seem to contain in a fairly stable membrane a number of enzymes capable of breaking down the unwanted different proteins of the cell itself including foreign particles ingested, the equivalent of the digestive system of the higher organism. Whether they act or not depends on the consistency of the lysosomal membrane which may be affected in different ways in health and disease and by drugs. A full understanding of intracellular biochemistry will clearly be the key to a much more rational medicine. Some organelles have an internal structure consisting largely of elaborately folded bimolecular lipoid membranes. The basic common structure seems to be an extensive system of folded membranes, the endoplastic reticulum, which separates two volumes of fluids – external plasma and internal cytoplasm proper. One part of the reticulum supports the ribosomes, another the Golgi apparatus, which seems to act as a digestive system for the cell. Our knowledge of the cell is now passing from the descriptive or Keplerian stage that it reached by the middle of the century to the interpretive or Newtonian stage which is clearly beginning with the working out of the nucleic-acid–protein synthesis. We are now beginning to relate what can be seen in the cell with what the cell actually does.

CELL DIVISION AND GROWTH

One of the most important parts of cytology is the detailed study of reproductive cells, of fertilization and the multiplication of cells to form a new organism. The interest in the growth of an animal from the egg has gone right back to the origins of science itself. Contrasting views on it were expressed in the eighteenth century by the *preformationists*, who believed the whole organism actually existed folded up in the egg, and the *epigeneticists*, who considered that every organism was made afresh through the action of a formative spirit.[6.208]

Another version of the same quarrel was taken up again towards the end of the nineteenth century between the *mechanists*, who wished to

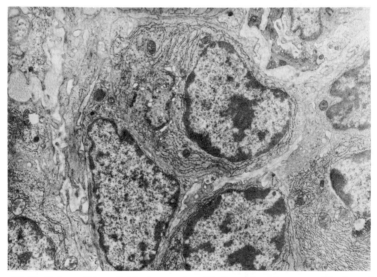

286. The living cell is an intricate organism. Photograph under a magnification of 12,650 times of a plasma cell in lymph tissue. Photographed at Guy's Hospital Medical School.

show that the growth of every individual was completely determined from the egg onwards, and the *vitalists*, who considered that every part of the egg had the potentiality of growing, through the influence of some formative agent, into the whole organism. The latter scored a success when Driesch (1867–1941) showed in 1891 that a sea urchin's egg divided in two gave rise to two complete larvae, not to two half larvae, but the mechanists scored when Loeb (1859–1924) showed in 1900 that it was possible to induce an unfertilized egg to produce a complete organism by subjecting it to chemical treatment. Some of these contradictions were removed when Spemann (1869–1941), Holtfreter, and Mangold demonstrated in 1931 that certain chemical or mechanical stimuli, when applied to an undifferentiated egg, were capable of inducing the formation of the organism as a whole; while others, acting only at a later stage when the organism had begun to grow, were capable of producing various parts of it, such as an eye or a limb and even supplementary eyes or limbs.[6,206] The nature of these *organizers* is still obscure. They may have some analogy with the sex hormones, which had been found to induce the secondary sexual characters at puberty, and indeed these changes may be considered as embryonic changes which are delayed until a much later stage in individual development.

These studies of chemical embryology seem to show that the general development of organisms must be controlled, as much as is their normal and abnormal metabolism, by chemical factors. The problem as to what determines the successive appearance of the different organizers or hormones at different stages of development is beginning to yield to the new interpretation in terms of successive liberation at different stages of coded information from the DNA of different cells. Here we may find a clue to the development for both the non-cyclic growth and the cyclic sexual changes.

It is known that the DNA as formed is often associated with small molecule proteins, the protamines or histones. These may well have the function of blocking whole sections of DNA and only liberating them one after the other. In this way the same DNA molecule can at one time stimulate the production of one protein and later on another. This, indeed, may provide the clue to the differentiation which occurs in the multicellular animals.

TISSUE AND ORGAN CULTURE

Throughout the century there has been an increasing drive towards an experimental study of growth and differentiation on all levels. From the study of the growth of eggs and embryos it has passed to that of higher organisms, with the mastery of the techniques of all tissue and organ culture by R. G. Harrison (1870–1959) in 1907 and by Fell in 1928. These studies have shown that, even after removal from the body, cells continue to grow and divide and for the most part retain their characters. Muscle cells remain muscle, bone cells grow as bone. There seems to be an internal regulating system of a chemical nature that controls the growth of cells in healthy animals, and prevents them getting in each other's way.

Later studies seem to show that in addition there is in some cancers a mechanism which prevents cells sticking together and thus from forming coherent tissues. These studies, particularly the following up of the early organ culture of Carrel (1873–1944), are already having important surgical applications and promise more. By careful attention to aseptic conditions and the development of mechanical suturing devices, organ-transplanting operations are becoming possible in animals and even in humans. By careful attention to these techniques, a human arm, severed in an accident, has been restored. A dog whose severed leg lay for two months in a refrigerator is now walking comfortably. Transplanted hearts and even mechanical hearts have been used successfully. This should before long do much to limit death and disability due to accidents or localized diseases.

CANCER

These studies are now undertaken largely under the impetus of the attempt to deal with the third group of diseases, those of unregulated growth. Under the general name of *cancer* they have become an increasing terror to mankind, particularly to those in industrial civilizations, where the greater average age exposes a larger proportion of the population to their attack. Now, the cancers differ from the other diseases in that they are, at least in their initial stages, strictly localized. They are diseases of cells which are transmitted from cell to cell and seen to spread throughout the body, mainly by the transport of cells, the multiplication of which forms the tumours characteristic of the disease. New cytological knowledge makes it clear that cancers are essentially diseases of the nucleus or, more specifically, of the nucleic acids that it contains.

Alteration in the heredity of a cell can be brought about in different ways, either directly chemical, by substances that can enter the nuclei, or through virus infections. The nucleic acid of the virus is enabled, so to speak, to contaminate the genetic material of the cell and thus modify its growth characteristics without actually starting a generalized virus infection. This seems to be the cause of the virus-borne jaw tumours that affect children throughout the low-lying grounds in West Africa, but are absent from the hills, where the insect that carries the virus cannot live.

A cure in the medical sense for cancer is not yet in sight, apart from such successful surgical intervention as removing the affected part when that is possible. Nevertheless, knowledge of the nature of cancers is advancing in such a way that it seems likely that some definite control may be achieved in the fairly near future; but only if research and

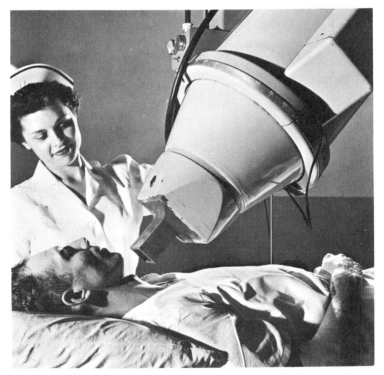

287. One form of cancer treatment is to inhibit the multiplication of malignant cells by irradiation with short-wave radiation. Originally gamma radiation was obtained by inserting radium needles but now artificial radioactive elements may be used. Radioactive cobalt emits gamma-rays and is here being applied to a patient. Photographed at Oak Ridge Hospital, Tennessee.

application of research on cancer are carried out in a far more vigorous, orderly, and scientific way than they are at present. It is of course very natural that with such a terrifying disease efforts should be directed primarily at curing and secondarily at understanding what is happening, but this is a short-sighted view. Control and understanding are equally essential. 'Practice without theory is blind, theory without practice is sterile.'

NEW LINES OF ATTACK

The first stage of the attack on cancer is the search for its origin and, correspondingly, for ways of preventing its onset. For a long time cancers have been known to be produced by certain chemicals: the original observation of John Hunter (1726–97) in the eighteenth century on the cancers of chimney-sweeps pointed to tar products, and these have been identified. Similar agents are suspected in the apparent connexion between smoking and cancer. The greater liability of heavy smokers to lung cancer now seems beyond doubt, but there may be other factors, such as diesel fumes, which also intervene. As yet no government has dared to take more than token steps in face of the unpopularity and loss of revenue which would follow from preventing people from killing themselves in this particularly unpleasant way. If cancer can be caused by chemicals it may well be cured by them. The real problem is to get the chemical to the place where it will be most effective against the cancer cells without damaging the healthy ones. A solution which has been used without much understanding for a long time is treatment with X-rays and radium. It now appears from the work of Lea, Bonét-Maury, Magat, and others that these radiations do not act directly, but by producing powerful chemical radicals such as OH which are more effective in attacking rapidly dividing cells than normal ones.[6.193] This is probably because the DNA is in a state of replication at that time and can be attacked by radicals.

Secondly, the compounds producing cancers are closely related to some of the hormones, particularly the sex hormones, which themselves produce cell multiplication; and one type of cancer at least, cancer of the prostate in man, has been actually cured by the application of such sex hormones. Here again there is a connexion with various kinds of virus disease already discussed (pp. 912 ff.). Thus the study of cancer is closely linked both with biochemistry, cytology, and with the study of viruses; and only by a very much greater and planned research effort in all these fields, without at first any relation to the cancer problem, is its solution confidently to be looked for.

## 11.7 The Organism as a Whole and Its Control Mechanisms

One of the major struggles between the mechanists and the vitalists which came to a head in the twentieth century was the concept of an organism as a whole. It is another aspect of the old conflict dating right back to Greek times between form and matter. The Pythagorean–Platonic view was that each organism, as an individual, must have a something corresponding to that individuality, a soul, psyche, or breath of life (pp. 178 f.). This is an old magical idea rationalized by the Greeks and transmitted by the Arabs into modern science. Those, like the primitive Buddhists, who saw no evidence for the existence of souls wished to find some other coherent reason for the unity and apparent purpose of an animal. The solution that naturally suggested itself in the Renaissance and was ardently championed by Descartes was that animals were machines. Men of course were different: they possessed a rational soul implanted by God (pp. 445 f.).

VITALISM AND MECHANISM

In modern science the difference between these two views was philosophically fundamental. Belief in a soul furnished an explanation of behaviour, satisfactory in itself, which did not require further research, since any action of the body as a whole was put down to the activities of the soul which, as spiritual, were beyond scientific investigation. To explain it without a soul, however, required a far more careful analysis of the operation of the body machine and called for experimental investigation. In practice, the difference was more apparent than real; the vitalists, although for their own sakes not requiring explanations, needed to study living organisms in order to show that the mechanists' interpretation of their workings was faulty, and were continually issuing challenges which acted as a powerful stimulus to further discoveries by the mechanist school. The fact was that in the seventeenth century, and almost to the end of the nineteenth, knowledge of the physiology of animals had not advanced far enough to provide any really rational explanation of how the animal worked as a whole, and therefore left the door open to the spiritist type of explanation (p. 398).

Twentieth-century research has gone far to provide a rational and material one. The maintenance of the vital functions of animals' respiration, digestion, excretion, had been considered by the Ancients as the concern of an inferior vegetative soul, in contrast to the nobler animal soul that directed outward movement (p. 223). Until the nineteenth century no better explanation could be offered, and even now much of the picture is still obscure. Nevertheless by observation and experiment much has been made clear.

### RESPIRATION AND DIGESTION

To the older impulse from medicine has been added in recent years a new drive, arising from the need to cope with the abnormal conditions which some individuals have to face in a mechanized and militarized world. The limits of resistance of the body to the pressures of deep diving and the anoxia of high flying or mountain climbing led to intensive research on the function of respiration, financed largely from State funds, concerned with the survival of trapped miners or of submarine and air crews. J. S. Haldane (1860–1936) and his son J. B. S. Haldane (1892–1964), by quantitative measurements and heroic experiments in which they were their own subjects,[6.186] explored the limits of human tolerance of different gas concentrations and provided a rational picture of how the body coped with the considerable variations which were tolerated. It proved to be so complex, involving lungs, heart, nerves, and brain, that J. S. Haldane was driven to accept a supernatural explanation, though his son found it equally compatible with materialism.

The study of digestion, carried on in a desultory way over the centuries, received two new impulses: one, already mentioned, from biochemistry, the other from experimental physiology. Biochemical methods resolved the successive breakdown of food materials by the enzymes of ptyalin, pepsin, and trypsin, the absorption of the products of digestion by the intestinal mucosa, and their later transformation and storage in the liver. These are all detailed chemical activities which can be studied in isolated preparations. For their co-ordination it is necessary to take the whole animal.

### PAVLOV

This is what Pavlov (1849–1936) did in 1897, inaugurating a new era in physiology. This was not just because he observed and experimented; here Pavlov was following Spallanzani (1729–99) and Beaumont (1785–1853). It was rather in planning and carrying out the new kind of systematic, quantitative, physiological investigation of which he was the pioneer. Pavlov's genius lay in his ability, while using experiment to find an answer to a specific question, to notice and follow up a side reaction. It was thus that he was led from a determination of the rate of secretion of gastric juice to the discovery of the conditioned reflex, of which more later. He established that digestion was no mere chemical cookery in the stomach, but a highly complex interaction of the whole animal to the stimuli from the stomach, the mouth, the nose, and the eyes, mediated by connexions with both the central and the sympathetic nervous systems. The unity of the organism is built into its structure, itself a product of long evolution.

Advances of the same kind, involving clinical, experimental, and biochemical studies, have been made on other bodily functions, only to reveal ever more complex interactions. This revelation of complexity is not a step backward, for each new discovery increases understanding and control. Thus, for instance, in his studies of the comparative biochemistry of excretion, Needham[6.208] has demonstrated the evolutionary sequence. Nitrogen is excreted as ammonia in simple water-living animals, where it can easily wash away. In most larger animals, including mammals, it is excreted as the relatively insoluble urea, which can be stored without damage to tissues. The final stage is the production, in reptiles and birds, of almost insoluble uric acid, which, he suggests, was evolved to save the very limited water supply available for their development inside eggs.

ENDOCRINOLOGY

The most significant of all the recent advances has been the study of the action of the endocrine organs – the ductless glands which produce the hormones, already discussed (p. 899). These glands are not isolated units; they respond themselves to other chemical and nerve stimuli; they are the chemical regulators of the whole body. They are concerned not only with normal maintenance and growth, but also with response to internal and external stimuli. One of the first of such actions to be observed was that of the hormone adrenalin, liberated under conditions of fright or anger, which stimulates the whole body to an effective response, flight in the one case, combat in the other.

Further research, particularly on the sex hormones, shows that the chemical control mechanisms are far more complicated. Each different hormone not only has its specific action, but it also reacts with other hormone-producing glands, and stimulates them to increase or to decrease their own hormone production. Indeed, there seems to be a general hormone or *endocrine* system, directed chemically from the pituitary gland at the base of the brain, which can send out separately some dozen different hormones affecting other glands in different parts of the body. Moreover the nervous and endocrine systems are in constant and complex interaction. In part this is manifest through the nervous connexion between the pituitary gland and the hypothalamus of the brain. Hormones affect emotions and in return emotions affect hormone production.

It would appear as if the body possesses two *communication systems* which duplicate each other, the slow postal system of chemical messages and the rapid telegraph of the nerves. The latter may be a secondary development, or both may have evolved side by side. In any case it is becoming apparent that the functional unity of the organism is not due to a simple mechanical juxtaposition of parts. The reason for invoking primitive directing entities, souls or entelechies (pp. 398 f.) was to try to account for its behaviour in terms of the hierarchy of society, effectively class society, as the fable of the revolt of the stomach in Shakespeare's *Coriolanus* shows us. Modern science cannot consistently make any use of such ideas; it must strive to unravel the structures and processes that secure the unitary functioning of the organism in its environment, and to account rationally for it in relation to its evolution.

288. Testing the nervous reactions and operational efficiency of an astronaut in a simulated space cabin, using an electrode head band. Photograph taken through a one-way window at the Lockheed Human Factors Research Laboratory near Atlanta, Georgia.

## THE ACTIVITY OF THE NERVOUS SYSTEM

So far we have spoken only of the relatively slow vegetative processes of organisms. In their immediate reaction to their environment the complex of sense organs, nervous system, and muscles are invoked. The study of this system reaches, as we have seen (p. 188), right back to the very origins of science, but twentieth-century researches have taken us a great step forward in understanding it. By the end of the nineteenth century the nervous system of man and of many different types of animals had been mapped anatomically, and its main connexions had been established by observation of local failures of movement and sensation associated with diseases or injuries to its various parts with the aid of animal experiments. It had been shown to consist in man and higher vertebrates of a central nervous system, stemming from the brain and responsible mainly for conscious sensation and voluntary movement, and less centralized sympathetic and parasympathetic systems, responsible for the considerable but unconscious movements and secretions of the inner organs.

## THE ELECTRICAL NATURE OF THE NERVE IMPULSE

Nevertheless how nerve messages were transmitted and how they were integrated were still very largely unknown at the beginning of the twentieth century. Without modern biophysical and biochemical methods it would indeed have been impossible to understand anything of the essential structures and processes involved. Adrian and others, using electronic amplifying systems, were able to demonstrate in the years after 1926 that the nerve signal consisted of pulses of electric potential all of the same strength but whose frequency, up to a definite limit, was proportional to the strength of the initial stimulus. Nerves therefore could transmit only information as to the quantity of an impulse, and its peculiar quality, such as colour, tone, or feeling, had to be inferred from the position of the channels along which the message was sent.

This analysis has had an enormous effect and will have an even greater one on our understanding of thought and consciousness. A large number of nerve messages never reach consciousness at all, but they are not unco-ordinated. Many are associated together in reflex arcs, where a certain sensation generates automatically a certain movement. One of the great achievements of the twentieth century was the work initiated by Pavlov in 1897, which shows that these reflexes were not entirely independent of the mind, but could be attached to each other and modified by consciousness. The experimental study of *conditioned reflexes* marks the highest level of the approach to psychological processes starting from physiology.

The movement of electrical action potential in a nerve impulse has more recently been shown by Buchthal, Hodgkin, and others to be essentially due to the propagation of a state of electrical polarization along a membrane through the transfer of metal ions from one side to the other. The operation of nerve impulses in generating movement or in receiving sensation has on the contrary been shown to be essentially chemical. As Dale and Dudley showed in 1929, at a nerve ending or at a synapse connecting two nerves the arrival of an electrical impulse liberates a chemical which in turn stimulates the cell which is to pass on the impulse.

NERVE CONNEXIONS AND ELECTRONIC SYSTEMS

It is through the interconnexion of nerve impulses, especially in the brain, that the study of biology is becoming more and more closely linked with that of advanced physics, particularly electronics. In 1928 Berger detected the passage of waves of electric potential between electrodes placed on the head of a patient. This led to the development of even more sensitive *electro-encephalographs*, which are of great value in the diagnosis and treatment of brain diseases such as epilepsy. It is also, in the hands of investigators like Grey Walter, beginning to throw light on the electrical occurrences that accompany perception and thought.*[6.142]

Much has been learned from the study of the brains of simpler creatures. The study by J. Z. Young of the brain of the octopus, the most intelligent of molluscs, is beginning to reveal something of the connective pattern which transforms the impulses derived from the sense organs into muscular contractions which determine movement.

Here it has been difficult to avoid the analogy between the brain and the servo-mechanisms and computers now being developed so rapidly by electronic engineers (p. 782). There we find three principal formal elements: the *coder*, which translates incoming messages into a form usable by the machine; the *machine* itself, including a *memory* for retaining information not immediately usable; and the *decoder*, which translates the machine messages into some external action. These correspond roughly to the *sense organs*, *brain*, and musculature or other *effector organs*. It would appear that the electronic computing machines themselves are extremely simplified versions or analogues of structures and active connexions already well established in the brains even of quite primitive animals, which are raised to a far higher degree

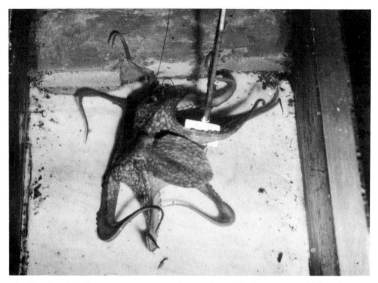

289. Study of the brain of the octopus by Professor J. Z. Young and his colleagues at University College, London, is revealing something of the connective pattern which transforms the impulses derived from sense organs into muscular con- tractions which determine movement. In this photograph, an octopus is making its first attack on a crab (suspended by a thread). The white plate gives the octopus an electric shock and in later reactions it learns to leave the prey when the white plate appears.

of complexity in the human brain. The study of the senses is beginning to show detailed analogies between the connecting systems that are shown by the microscope in the brain and those that are built up in various kinds of circuit. The visual cortex, for instance, appears to be constructed in layers of which the first is concerned with the actual analysis of the pattern presented by the retina of the eye and the others, concerned with interpreting it, turning the primitive percept into already a fairly analysed concept such, for instance, as that for a mov- ing object. Despite the fact that its position, colour, and shape may change, it is thus recognized as an entity. We are still very far from being able to construct a mechanism that can approach the brain in its complexity. The essential difference is that the brain is a highly minia- turized system in which the slowness of the individual acts of stimu- lation and transmission are compensated by the far higher order of numbers in the reacting cells, something of the order of ten billion.

This does not mean that the brain is a computing machine, any more than the eye is a camera. It does mean that we can learn much from the analogies between natural systems which have evolved over a long period of time, and have acquired a greater perception and more accurate analysis of their environment and consequently a greater control over it, and artificial systems deliberately designed to extend man's capacities in both these directions.[6.142]

### ANIMAL BEHAVIOUR

A different approach to the problems of the inner co-ordination of animals was developed largely in the twentieth century by the scientific study of animal behaviour. Thanks mainly to Pavlov, this has been brought into relation with the study of nervous mechanisms. The behaviour of animals has been studied by man for an enormously long period, perhaps most intently in the Old Stone Age, when man was beginning to hunt them, and at the beginning of the New Stone Age, when they were being tamed for the first time. After then such knowledge became traditional and interest in new aspects declined. From being utilitarian, it became first magical, just one out of many devices for telling the future, and then morally edifying, as in the animal fables of classical times and the bestiaries of the Middle Ages with their courageous lions, sly foxes, and self-sacrificing pelicans. Finally in the Victorian age it became merely anecdotal and sentimental, concerned mainly with pets and with sport. A serious quantitative study was late in developing largely because it was thought that there was nothing to explain. By definition animals which lacked *reason* must do everything by *instinct*. In reality a new world was waiting to be revealed. Darwin, here also, was a pioneer with his studies on *Expression of the Emotions in Man and Animals*.[6.168]

### INSTINCT AND LEARNING

The experimental study of animal behaviour started with the work of C. L. Morgan (1852–1936) on the way in which animals, ranging from chickens and rats to monkeys, reacted to certain situations and tried to solve certain problems. The difficulty was to find situations sufficiently similar to those the animal was used to and sufficiently simple to control and interpret. Early results, especially those of Watson (1878–1958) in America and Köhler in Germany, seemed to reflect more the mind of the investigators than those of the animals. The one found completely random, the other thoughtful, behaviour. On these they constructed two radically different theories; Watson left out the mind altogether,

290. Instinctive animal behaviour as shown by two Impala rams parrying with their horns during the mating season. Photograph taken in Kruger National Park.

followed the pragmatic line, and claimed that all that existed for men as well as animals was stimulus and appropriate behaviour; Köhler peopled the mind of animals with new unitary constructs – the Gestalts – curiously recalling Plato's ideas.

In more recent and critical work a new beginning has been made in exploring the field of interest of various animals and something of the workings of the most significant of mental abilities: the capacity to learn by experience. *Learning* implies memory, but is something much more advanced. Experiences received have to be stored, compared, and selected, and new experiences have to be made before a learned pattern of behaviour can be established. This concept of learning has, however, already been introduced into electronic machines, though on a very simplified scale, and can lead to the construction of automatons that can move around on their own and reponds in a 'learned' way to the situation around them.

MEMORY

The nature of animal memory is now being studied by a variety of experimental methods. It would appear that memory is a very early character of organisms. May's experiments with such lowly creatures as flatworms show that the pattern of response is transmitted to the whole body. Flatworms cut in pieces and regenerating seem to have some memory of their previous experience. More recent experiments on mice give some indication that the memory may be actually stored in altered varieties of nucleic acids in the sense organs, possibly also in the brain. If this were confirmed it would seem to show that memory is more like a set of photographic images printed out, so to speak, not in silver but in nucleic acid capable of being referred to over and over again through appropriate stimulation. The intermediate idea that the memory and nerve system might consist of continuous repetition of the same message circulating round in an active way may have to be – at least in part – discarded. There is no reason to believe, however, that there is only one kind of memory; there may be both short-term and long-term memory.

## THE LANGUAGE OF ANIMALS

By a combination of careful observation and subtly planned experiment, it is now becoming possible to interpret the behaviour of animals in the state of Nature, including their mating, their care for their young, their relation to other members of their own or of different species. Such studies are indeed taking the place of the old anecdotal natural history and satisfying the age-old wish of learning the language of animals, which turns out to be, though simple in comparison with ours, still of great complexity. What is appearing from the work of such observers as Tinbergen[6.227] is that behaviour patterns needed for moving or feeding can be displaced, modified, or exaggerated to convey meaning and impose appropriate behaviour on other animals. Thus the birds' song can call mates or warn off rivals. Even the bees, as Von Frisch's beautiful studies have shown, have their own language expressed in dancing for indicating the direction and distances of sources of honey. Social animals, as might have been expected, nearly all have some kind of language. Some of these are mediated not by sight or sound, but by smell, as with the pheromones or special sense used by ants to indicate the presence of food. Such studies are of the most profound importance, not only because of the light they throw on nervous mechanisms, but also on the origin and nature of human communication and the society it forms and binds together (p. 1028). For the same reason attempts to interpret animal behaviour, even the simplest, meet difficulties which do not lie entirely in its logical complexity. In this boundary region it is difficult to eliminate relics of human thought and language, the more so because, as will be argued in the next chapter, they are tied to religious and political prejudice.

DISEASES OF THE MIND: PSYCHOLOGY AND PSYCHIATRY

These considerations apply in still greater force to progress in the last and most difficult branch of medicine – that relating to mental disease, which may manifest itself in both mental and bodily form. Here the last fifty years have witnessed an enormous interest but no secure progress. It has been a field of controversy and fashion. Many would claim that the work of Freud was as epoch-making as that of Einstein in the twentieth century. It certainly proved to be so in the negative sense of clearing away much philosophic rubbish from psychology. What was put in its place, however, were simply new *ad hoc* word constructs – the unconscious, the id, complexes, and repressions – which have now passed into current usage among the intellectuals of capitalist countries (p. 1154). This metaphysical basis for psychology has not been justified by any secure experimental evidence. In curing mental ills, Freudian psychology, in its original form or any of its variants, has not lived up to the early hopes placed on it, though it has proved soothing to those who can afford to pay for the treatment. The greater appreciation of the extreme complexity of the brain is, however, discouraging the alternative ways of dealing with nervous disorders by violent electrical or surgical interference, shock treatments, lobectomies, etc. On the other hand, it seems that one of the commonest of mental diseases, schizophrenia, seems to be partly of biochemical origin although its onset and progress may be determined by psychological states. An enormous amount of research will be required before these interactions of nerve currents and biochemistry are sorted out.

The human mind is neither on the one hand merely a set of nerve connexions, nor on the other a set of disembodied entelechies – spirits, instincts, complexes. It is the means that man has evolved, not by himself but in society, for dealing with his environment, itself increasingly a social one. For that reason psychology divorced from the economic and political basis in society is bound to be on the wrong track. The alternative way, because it is essentially a matter for social science, will be dealt with in the next chapter.

## 11.8 Heredity and Evolution

The remaining sectors of biological advance in the twentieth century – those of heredity, evolution, and ecology – are more closely connected with agriculture than with medicine. In the nineteenth century interest in the wide range of animal and plant species was still largely that of the collector and naturalist, concerned in the first place with making an inventory of the world of life, *extensively*, by far-flung explorations, and *intensively* by the use of the microscope. Even the explosive effect of Darwinian evolution did not at once change this trend. Its first consequence was to canalize biological effort along the same naturalistic lines in an attempt to put some order into the collection, to provide life with its family tree. Meanwhile practical agriculture pursued its traditional way, with the help of science only in farm machinery, fertilizers, and a few animal medicines. Improved breeding was left to practical men and enthusiastic amateurs and fanciers.

THE DRIVE FOR BETTER BREEDS OF PLANTS AND ANIMALS

Inevitably, however, at the turn of the century, with the Darwinian controversy dying down, came a new need for science in practical breeding and a corresponding interest in heredity. By then it was apparent that the opening up of the Western Prairies for wheat, of Australia for sheep, and of the new empires carved out of Africa and Asia for tropical products, emphatically required something better than what the old hit-and-miss breeding would provide. On the scientific side, the main lines of evolution having been laid down, interest began to centre on its mechanism. The laws of inheritance began to acquire a new importance. Now even though we do not understand how it is that an egg or a seed will grow into a frog or an oak tree, we can still be reasonably sure that they will do so.

The likeness of offspring to parents is not however absolute; some of the offspring are usually bigger or in some way more useful to man than others, and these have from before the dawn of history been used for breeding. It would have been absurd to wait for the complete explanation of the phenomena of heredity before formulating provisional laws by which we could learn to control them. This has certainly been the case in the history of genetics, which developed very late into an independent branch of science and one where practice has long preceded theory. Deliberate breeding must have started with the very dawn of agriculture and animal husbandry. Pedigree horse-breeding, for instance, is known from documents as far back as 2000 B.C., and indeed the main methods of transforming species for practical use or for sporting taste have been well established, if only on a purely empirical basis, from the early days of civilization and even earlier in the Old Stone Age with the domestication and variation of the types of dog.

### DARWIN AND VARIATION

Because it is not only animals that breed but also human beings, and because the establishment of class and race distinctions in society turns on inheritance, the problems of genetics have been, and still are, continuously bedevilled by religious and political considerations. The very word inheritance or heredity is bound up with the essentially social concept of passing on property to an heir. Its use suggests that even in biological inheritance something material or formal is handed on – like the Habsburg chin with the title to the Holy Roman Empire.

291. The effect of environment on evolution is well illustrated in the case of the Peppered Moth (*Biston betularia*). In the photograph it can be seen in its normal form (above) and in its melanic form *carbonaria* (below). The moths are resting on a soot-covered tree near Birmingham.

Darwin's theory of evolution focused great attention on the principles of the variation of inheritance, but actually raised far more difficulties than it solved. He thought that species might vary in response to environment and that selection would operate on those variations. He was, to a certain extent, a Lamarckian, in the sense that he thought that environment directly influenced this variation. The marvellous adaptations of organisms to the most varied conditions of environment all pointed to some such moulding effect. Darwin's ideas on heredity, however, were not immediately fruitful, largely because they had no quantitative experimental basis. It seemed impossible to show in practice how variations occurred and how they were fixed (p. 644).

HEREDITY, CLASS, AND RACE

Though Darwin himself had drawn much of his material from stock-breeders and fanciers, his successors, who were more academic, lost touch with them. Belief in the importance of breeding was indeed an old and an essential support of the aristocratic feudal system, but until the nineteenth century it had not called for scientific justification for it. When support was needed against the radical tide of the latter nineteenth century, science was called in under the banner of the eugenics movement. Its early success was largely due to Galton (1822–1911) – a wealthy, well-connected amateur, incidentally Darwin's cousin – and Karl Pearson, a mathematician and positivist philosopher, almost the first to apply mathematics to biological problems (pp. 1085 f.). Both were concerned essentially in justifying on scientific grounds the moral position of the middle and upper classes, which was beginning to be shaken by equalitarian socialist agitation, by proving that they were genetically superior to the lower classes. In baser hands the same arguments could be used to prove that the white races were superior to the coloured, or the Nordics to the other white races, particularly to the Jews (pp. 1086 f.).

WEISMANN AND THE PERSISTENCE OF THE GERM PLASM

The fixity of species and inheritance was further emphasized at the end of the century by Weismann (1834–1914), who, on the basis of repeated failures to produce any inheritance of acquired character, developed the formal theory of the continuity of germ plasm – a kind of family treasure handed on undiminished from parents to children, and suffering modification only by the mixing inevitably produced by sexual reproduction. In this view the living organism or *phenotype* was but one of many fleeting expressions of the perennial *genotype*. This was, in the nineteenth century, almost a full return to the ideas of the seventeenth-century preformationism. It effectively made nonsense of evolution, as it implied that the potential characters of all animals and plants were present in the first germ and only need to be sorted out. Further support for the all-importance of inheritance came from the field experiments of Vilmorin (1816-60) in 1856 and of Johannsen (1857-1927) in 1903; they showed that ordinary crops consisted of individual plants of very varied heredity, but that by careful inbreeding and selection it was possible to produce pure lines which, in principle, would continue to breed true for all time.

THE DISCONTINUITY OF HEREDITY:
THE REDISCOVERY OF MENDEL'S LAWS

As long, however, as the variations in inheritance were considered to be continuous all this work remained necessarily purely descriptive, and could not be connected to the rest of biology. The recognition of the existence of discontinuous changes transformed this position. Bateson (1861-1926) in 1894 claimed that it was these sharp variations rather than indefinite graduation that was significant in evolution.[6.105] In 1901 de Vries (1848-1935) discovered abrupt changes – *mutations* – among evening primroses. Both found support in the experiments, made between 1857 and 1868, and laws of Mendel (1822-84), published in 1869, which had been neglected in his own time and which they re-discovered and extended. Mendel, working with peas in his monastery garden in Brno in Czechoslovakia, had shown that many characters were transmitted in sexual inheritance in a peculiarly simple way, which he interpreted as indicating the existence of certain unit factors deter-mining such things as the colours of flowers or the wrinkliness of seeds. The great initial advantage of this *gene* or unit theory of inheritance was that it was essentially simple and mathematical. But there was, of course, the danger, which was not appreciated at once, that study might be confined to those of characters that did show these simple relations and that a theory that explained some part of inheritance should be thought of as explaining the whole of it.

MORGAN: GENES AND CHROMOSOMES

The simple laws of Mendel seemed all the more plausible when a con-nexion was established between the unit genetic characters and the chromosomes that had been observed in the nuclei of dividing cells. This was essentially the work of T. H. Morgan in America. Beginning in 1910 he made an extensive study of the whole range of variation of

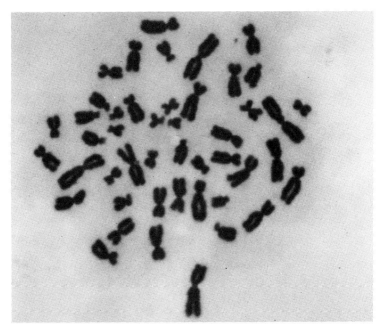

292. Chromosomes photographed in normal male lymphocyte. The chromosomes can here be seen in pairs.

one small fly, *Drosophila melanogaster*, which had the advantage of breeding very quickly and being very easy to keep. The simplicity and exact mathematical character of the genetic theory encouraged an enormous amount of research relating the detailed characters of the fly to its chromosome structure.

This led to the discovery that various characters that were often inherited together could be associated with certain parts of a chromosome, also lying close together; in other words, that the chromosome corresponded to a map of the whole development of the characters arranged in line along it. It was accordingly assumed that to each inherited character that appeared in the adult organism there corresponded a material particle, the *gene*, in one of the chromosomes of its parents. Each cell of every organism contains a set of pairs of chromosomes, one derived from each parent, and should therefore possess a pair of genes for every character. The process of breeding consequently reduced itself to different ways of shuffling and dealing out the genes of the offspring. If any character of one parent did not appear in the offspring it was assumed that the gene was absent or suppressed by a stronger or dominant gene from the other parent. Although the genes were not supposed to be material bodies, this was really an inference from their location in the chromosome.

MUTATIONS NATURAL AND ARTIFICIAL

As time went on the genetics of *Drosophila* and a few other organisms were worked out in great detail. The genes did not appear to be entirely stable. Already, in 1900, de Vries had observed that new characters would appear occasionally and without warning, even in inbred lines, and would breed true. Now the occurrence of these *mutations* suggested that the genes were subject to chance variation, and that their appearance could be affected by external circumstances. This was confirmed by Muller, who in 1927 showed the increased production of mutants by means of X-rays. Since then it has been shown that other agents, such as specific chemicals like colchicine, also produce mutations. These observations, somewhat academic or at most of agronomic interest at the time, have since become of the most vital importance to humanity because of the production of such mutations in plants, animals, and human beings, through the effect of radiation produced by atomic and hydrogen bombs and their radioactive fall-out (pp. 954 f.). Already a considerable number of such mutations must have been produced in Japan in 1945 and subsequently, though far more widely diffused as a result of bomb trials (p. 840).[6.184; 6.188; 6.204; 6.205; 6.231]

EFFECTS OF RADIATION ON ORGANISMS

The alarm caused particularly by the effects of the hydrogen bomb brought to light the very limited knowledge available on the effects of ionizing radiations of all kinds on organisms, and active research in this field is now proceeding. Heavy doses of radiation, such as those produced by accident in operating nuclear piles or in direct exposure to bombs or fall-out, produce violent reactions in all affected cells, probably due to a general upset of cell metabolism by larger numbers of OH radicles interfering with enzyme reactions (p. 892). At lower, but still severe, radiation levels the main damage is to the nucleic-acid-protein synthesis mechanism and therefore affects primarily dividing cells, producing permanent damage or starting a process of cancer formation. One group of cells attacked are those which produce the white corpuscles leading to fatal leukaemia, often only after several years as has been recorded in Hiroshima.

293. Mutation in carnation caused by gamma radiation. The plant produces white flowers under normal conditions but after irradiation a red carnation appeared on the middle stalk. Photograph by the Brookhaven National Laboratory, New York.

## GENETIC CONSEQUENCES OF ATOMIC WARFARE

At lower doses still only a minimal change is produced in normal dividing cells, but in the special case of a germ cell the result may be a mutation which, if recessive, may not show up in inheritance unless the person carrying it is mated with another individual with the same mutation, which will happen sooner or later according to the concentration of the defect in the population and the degree of inbreeding. The effect of bomb-produced radiation at different levels can be estimated, but only very roughly, owing to inadequately supported research. In the enormous increase in strength and numbers of nuclear explosions to be expected in any general war, it would appear that the direct loss of life from the explosions themselves and their associated firestorms would predominate over major genetic effects, but, as there would probably be survivors on a small scale from such a war, their descendants, if any, would inevitably suffer from severe genetic effects. In the first phase there would be a predominance of abortions and monsters. In the reduced population that would succeed them defects due to recessive gene combinations would appear in a large proportion of births for hundreds of generations, long after detectable traces of the bombs' radiations had disappeared from the world. Nor, of course, would damage be confined to humans. Monstrous productions of animals and plants would also be inevitable, and it would take many thousands of years to reach any genetic equilibrium.

The effects of experimental explosions at the 1962 rate were distressing enough. Official reports, expressing results in percentage increase of mutations, heavily played down, may have seemed reassuring, but the radioactive dust is still settling from the upper atmosphere, and it will be at its worst in 1970.[6.188; 6.205] Already it is certain that tens of thousands of abortions, monsters, and defectives will result in ensuing generations, causing misery in as many families. The case for stopping this criminal folly before it became still worse seemed overwhelming, but it was not till 1963 that popular pressure on governments secured the signing of a partial test ban treaty which has not yet, after more than a year, put a stop to underground tests and the further development, let alone the stockpiling, of nuclear weapons. The nuclear threat still hangs over us and continually increases.

APPLICATIONS OF GENETICS

From the outset the gene theory of heredity found its chief interest, as well as its chief applications, in those cases where single gene effects were clearly marked and where the effects of different genes could be studied in different combinations. This occurs where the gene controls a single biochemical reaction modifying – or not modifying – some final produce as, for instance, in the colours of flowers, feathers, or urine.

These may or may not be physiologically important – in man, for instance, they may produce certain forms of idiocy – but they are in any case most useful as markers. The working out of the human genetic map by such workers as Haldane and Penrose is of considerable clinical value. In particular, the genetic determination of blood groups (pp. 901 f.) has been a means of saving the lives of many babies by indicating the necessity of blood transfusion. In agriculture the successes have been more limited. Certain disease-resisting types of plants have been produced following exact genetic practice, but most of the successes of genetics in improving, for instance, hybrid maize or cotton

294. Genetic change in sheep. The ram on the right is an Ancon (short-legged) and if bred with a normal ewe (left), it produces a short-legged lamb (centre). In the wild state this mutation would be disadvantageous, but it is useful to farmers since short-legged sheep cannot jump fences and this variety is bred specially.

have been got by much more rough-and-ready methods. The reason for this is that most of the characters of an animal or plant which are economically useful, such as weight or yield, must depend on a large number of genetic factors. Theory is not sufficiently advanced to deal with such complex cases, while complete genetic analysis would cost more money than a genetic department can usually raise, and the farmer cannot wait till it is completed.

### ALTERNATIVE METHODS OF MODIFYING INHERITANCE

The attention that has been given in recent years to the complex molecular mechanisms of reproduction in the DNA-RNA-protein cycle has tended to obscure the fact that these are not the only elements that can be responsible for the reproduction of organisms. Modern studies in cytology have shown the extreme complexity of the cell and we can by no means be sure that all forms of inheritance start in the DNA of the nucleus. The evidence of the virus alone shows that the RNA can act perfectly well to ensure regular reproduction of protein structures. The great number of organelles that are found in cells show that their method of reproduction must be complex indeed. Moreover, it is unlikely that it will be easy to reduce it to more or less fixed rules as with those of the DNA cycle.

It would be absurd, however, to wait for the unravelling of these mechanisms and to give up attempts to alter heredity by more empirical approaches to its control by varying environmental factors. This apparent contradiction, which is essentially one of emphasis, was the basis in science for the opposition in the Soviet Union, for instance, of the school of I. V. Michurin (1855–1935) and T. D. Lysenko, to Mendelian genetics. That controversy has now died down in the light of later understanding. Much work done in countries outside the Soviet Union or other Socialist countries has shown that there are numerous phenomena of inheritance, either by direct modification of environment as in flax or through grafting, which bear out some of the claims of the Michurin–Lysenko school, but up to the present no full explanation of their mechanism has been found. The physiological aspect of control of plant growth is now a little better understood and the effect of vernalization, that is, of change of temperature on various stages of plant growth, is beginning to be used well outside the Soviet Union.

In earlier editions of this book I have tried to give a more detailed account of the controversy, but it seems that now there is little point in doing so because the genetic approach to heredity has become fairly uniform all over the world. At its height the contestants were mostly talking at cross-purposes. The orthodox geneticists were concerned with understanding the mechanism and working out the consequences of an illuminating theory; the Soviet agrobiologists were trying to find the quickest way of improving the livestocks of their country, not so much by breeding and selection but by intelligent modification of the environment. This essentially depends on physiology and even more widely on ecology. The questions of heredity are, so to speak, embedded in these sciences.

EVOLUTION

The question of *evolution* is so closely linked with that of heredity that it seems more appropriate to treat it here, though logically it should come after a discussion of the interrelatedness of animals and plants, which is in theory *ecology* and in practice agriculture. In Chapter 9 the outlines of the great nineteenth-century controversy on organic evolution have already been drawn.

The triumph of Darwin was not so much in his discovery of evolution as in making it a scientifically plausible idea. Thanks to him, the fact that evolution did occur and is occurring is accepted by all but a few bigots. The end of the nineteenth century was largely taken up with determining the most probable chain of relationship between the different forms of animals and plants – in drawing up, so to speak, the family tree of evolution. In the twentieth century the interest shifted to establishing the mode of evolution, how and why new forms came into existence, when and where they did. Here no finality has been reached; in fact it is in answering such questions that the major differences of opinion in biological science are revealed.

NEO-MENDELIAN EVOLUTION BY MUTATION
AND NATURAL SELECTION

The last few decades have witnessed the rewriting of the Darwinian theory of natural selection in terms of the Mendelian theory of genes. Instead of the imperceptible variations postulated by Darwin, there have been put the abrupt changes produced by gene substitution, by gene multiplication, by chromosome doubling or polyploidy, and by gene mutation. All these changes are supposed to occur for reasons which have nothing to do with the adaptive value of the resultant character in the adult animal. On this view selection would operate, not on the characters, but on the genes or gene combinations which carried them. Natural selection would act as a kind of sieve, changing in a mathematically determinable way the gene composition of the population. Adaptation would simply be the most successful of a number of absolutely random shots. Formally, through the statistical labours of such mathematical biologists as Fisher (1890–1963) and Haldane, such a mechanism would account for the evolution of a species and even for its splitting into two non-interbreeding species, that is, for the creation of a new species. It is, however, extremely hard to test in practice, owing to the difficulty of controlling environments quantitatively, and even more to the slow rate at which evolution takes place.

INHERITANCE OF ACQUIRED CHARACTERS

Some experiments have been carried out in recent years with a bearing on evolution which in themselves may help to resolve the major difficulties still felt by naturalists as to whether simple selection of random variations is sufficient to explain the actual process of evolution. Ever since the time of Lamarck (p. 640) it has been felt that environment must in some way direct the course of changed inheritance to such obvious advantage of the organism. Yet very effort to prove inheritance of acquired characteristics has failed (pp. 950 f.). We now see why, in the case of higher animals and plants with their most inaccessible reproductive mechanism and their long reproductive periods, this should be so difficult. When we come to the lower organisms, such as bacteria, with life-cycles measured in minutes, it is possible to observe changes in response to chemical environment which are definitely adaptive, in the sense of being able to grow in new media. It would appear that here direct effects on the nucleo-protein hereditary material are possible, as was first shown as far back as 1946, when Avery produced a change from one type of pneumococcus to another by the addition of nucleic acids of the second type. We now know very similar effects produced by virus infections (pp. 934 f.).

This, however, is still a long way from adaptive inheritance connected with general environmental stimuli. One type of this has been discovered by Waddington. A certain strain of flies, when exposed as pupae to higher temperatures, develop wings with one bar missing, an apparently trivial variation. If the process is continued for several generations, however, flies are produced which breed barless even when the pupae have not been heated. After more generations without heating these revert to the normal form. Whatever the explanation, and Waddington offers a complicated Mendelian one, it would appear at least that imposed developmental changes may somehow be incorporated in the heredity of an organism in relatively few generations.

Even more striking cases are reported[6.182; 6.187] where habits and not structures are inherited. The wild canary was chosen some three hundred years ago as a cage bird on account of its plumage alone. It chirped but did not sing. However, bird fanciers in different places managed to teach them different songs, and now each breed of canary, even when brought up from the egg without teaching, sings in its characteristic style. Something of the same kind seems to have occurred in our own inheritance. It is not long since we have learnt to talk – 200 to 1,000 generations according to different estimates. Even now we have to be taught to speak. However, the human brain and ear have been profoundly modified so that a considerable area of cortex has been set aside solely for that purpose, and that without much evidence of selection against dumb types. Until we know something of the mechanism by which such rapid evolutionary changes can be carried out this kind of inheritance cannot be formulated as a theory, but it remains a warning against dogmatic exclusion and reliance on only one form of transmission of characters in evolution.

Not only is there the greatest theoretical interest in this question – the central one of heredity and evolution – but its solution will lead to important practical consequences. Breeding and selection are necessarily, especially in slow-breeding species, very slow processes; directed changes would enormously increase the rate of improving them, compared to the present hit-and-miss method of practical breeders. It is still quite premature to claim, however, that any methods of directing evolution have been found or are even in sight. Only intense and critical research carried over into expensive and planned practice can hope to discover them. Yet it is not intrinsically forbidden for man to hope to do consciously what has been happening in Nature since the origin of life.

## 11.9 Organisms and Their Environment: Ecology

The study of organisms in relation to a variety of environments, natural and experimental, is one that has grown rapidly in twentieth-century biology. Before then the account of an animal or plant was apt to be limited to a morphological, anatomical description, together with some physiological study of its separate functions and a natural history account of its habits. This knowledge is now felt to be only a necessary first step to the understanding of the far more complex and dynamic aspects of the life of the organism. Mere observation and natural history are not enough: detailed and large-scale experiments are also required.

As was pointed out at the beginning of the chapter, the century has witnessed the rise of a school of experimental biology and of the studying of the functions of living animals or plants by varying their conditions of life and observing the resulting changes. The place of experimental biology in the twentieth century is analogous to that of organic chemistry in the nineteenth. It is a way of finding out about the effective structure of organisms by studying their reactions to different environments, just as the chemist found the structure of his molecules by exposing them to different reagents. This cannot be done by limiting observations to the conditions in which the organism ordinarily lives. A wider range of possible environments needs to be explored. In more complex situations it is necessary to make a very careful analysis of all distinguishable factors in the environment, to vary them either one at a time or several at a time in a way determined by statistical methods, and to make equally complex observations on the organism.

INTERACTIONS OF ORGANISMS

The problem is made even more difficult by the fact that the environment of any organism invariably comprises innumerable other organisms. Darwin himself fully realized this in the middle of the nineteenth century, particularly in his work on the fertilization of flowers[6.167] and on the earthworm.[6.169] But the work that has been done up till now only emphasizes the extreme complexity of the relations between organisms and our practically total ignorance of their significance. The soil, for example, which is the basis of all plant and therefore of all animal life on land, is very largely a realm unknown to biological science, although it probably contains more living organisms in its bulk than are to be found on its surface. Until recently soil science has been essentially

295. The introduction of wild rabbits into Australia in 1859 brought about a disturbance of the balance of nature and special steps for eradication and limitation have been taken. The photograph shows rabbit-infested country in south-eastern Australia (left) and country free from rabbits (right) due to the construction of special rabbit-proof fencing.

descriptive and largely inorganic, based on geology and mineralogy. We are only now beginning to realize that the soil itself is a whole complex of organisms, not one of which can be changed without affecting all the rest.

### MUTUAL DEPENDENCE OF GROUPS OF ORGANISMS

The interrelated complex of organisms – of animals, plants, and bacteria – wherever they are found, is the subject of the study of *ecology*: the analysis of the total effect of all organisms in a specific locality on each other. The *association* of organisms in, for example, a field or a pond is found to have a coherence and permanency of its own greater than those of any individual organism. The old crude concept of the struggle for existence is giving place to one of the evolved co-operation of different organisms. The co-operation may sometimes take rather paradoxical forms; that, for instance, between carnivores and herbivores. The condition of deer is, to a large extent, determined by the degree to which they are thinned off or kept in condition by wolves or hunters, as the sad consequences of their introduction into New Zealand, without these predators, bear witness. But, by and large, a certain equilibrium is maintained in any uniform environment in Nature. No individual species can multiply, very much less die out, without affecting all the others.

A crude misunderstanding of the Darwinian phrase of the struggle for existence further obscures the real dependence of organisms on each other. It would be of little value to any individual or species to flourish at the cost of exterminating all others. This applies even more strongly inside a species than between organisms of different species. Nevertheless the concept of the struggle for survival is still in vogue, largely because it has been and is still so useful in justifying ruthless competition and the rule of the stronger in human affairs. As Lysenko has pointed out, it is only in exceptional densities of overcrowding, rarely found in Nature, that individuals of a species come into competition. For the most part, in plants as well as animals, the presence of other members of the species improves the adaptive character of the environment. A forest, for instance, is of net value to all the trees living in it.

### MAN'S INTERFERENCE WITH THE BALANCE OF NATURE

A new phase in the history of our planet opened when man began to interfere with the previously established balance of Nature in a way essentially different from that of any other organism. As a hunter, even more as a peasant – at first unconsciously and on a small scale, later

296. The Dodo (*Didus ineptus*), a native of Mauritius, had a large cumbersome body and small wings that were useless for flight. Unlimited hunting rendered the species extinct.

consciously and on a scale ranging over the whole planet – man set himself to tip the balance of Nature in his own favour. How successful he was at that from the start is shown by the multiplication and spread of the human race, which have proceeded with ever-increasing momentum. In the early stages man lacked an adequate understanding of what he was doing, and produced occasional undesired results, such as exterminating the game on which he lived and over-grazing pastures or exhausting patches of cultivation; but the small scale of these operations prevented any permanent damage to the earth's resources. Now the situation is different; neither knowledge nor power is lacking, but the success of modern mechanical agriculture and lumbering has been at the expense of ruining a dangerously large proportion of the soil of the planet and of changing its climate unfavourably to almost all forms of life.

DESTRUCTIVE EFFECTS OF AGRICULTURE UNDER CAPITALISM

This large-scale intensive devastation has nothing to do with the inherent wickedness or stupidity of man, or of his unrestrained desire to propagate, as many publicists want people to think; it is simply due to the essentially predatory nature of capitalism, now spread as imperialism over so large an area of the world. The destruction of the soil has been enormously accelerated in the last fifty years by the methods characteristic of ruthless capitalist exploitation for immediate profit. The actual destroyers of the soil need not themselves be capitalists, they may be poor share-cropping farmers who have to secure a large harvest of cash crops in order to prevent themselves being evicted; or Africans driven on to reserves by Europeans who take all the best land. The different causes lead to the same result, and the process is continually accelerating. The less there is in the land, the more it has to be exploited and the worse its condition gets.[6.161; 6.163; 6.171]

CONSERVATION

Nevertheless there is no absolute reason why this waste and destruction of irreplaceable resources should go on. Even under capitalism, spasmodic and limited attempts to check it show that this is technically perfectly possible. The American depression of the nineteen thirties gave us the Tennessee Valley Authority and a widespread movement of soil conservation; both were successful as far as biological engineering is concerned (pp. 1020 f.). But private interest has seen to it that the TVA should remain a lonely specimen of what can be done in regional planning, and the soil-conservation measures are limited to where they can be afforded and are abandoned whenever intensive farming becomes profitable.

This whole problem, however, now is wearing a new aspect because so many of the former colonial countries are gaining independence. This independence, however, is not a genuine one as long as the economy of the country still depends on mono-culture of cash crops, either as plantations or for sale to the agents of the industrial capitalist countries. The examples of the socialist countries in the tropical zone, particularly of China, and now of Cuba, have shown that it is possible

297. Soil conservation and irrigation can transform barren areas. In the Sahara near Tourgourt, discovery of deep artesian wells and the construction of windbreaks and irrigation channels has made it possible to cultivate a palm grove.

to proceed simultaneously with the improvement of the agricultural yields for the benefit of the population and a conservation policy for the land.

### TRANSFORMING NATURE

In the part of the world now saved from the operation of the free market and the monopoly trust the picture is a very different one. There, more especially in the Soviet Union, where alone there has been time for long-term projects to mature, the improvement of soils and the reclaiming of deserts have been going on for twenty years. The land there belongs to the people, and its permanent preservation and betterment are a first charge on capital investment.

298. Construction of a reinforced concrete conduit for an irrigation system in the arid area of the Tajik section of the USSR.

In post-war years this process has been so extended and accelerated as to constitute something radically new in the history of our planet: a deliberate attempt to remake Nature and change geography in the service of mankind. To be able to imagine and carry out such an enterprise requires, in the first place, a people accustomed to working for common enterprises, and confident enough to make sacrifices in the present to secure greater rewards later. To make this goodwill effective, however, requires the greatest use of science: of mechanical engineering to dam rivers, cut canals, and build power-stations, of biological engineering to plant forest belts, to install irrigation, to balance animals and crops.

Very considerable steps have already been taken to safeguard the drought-threatened and semi-desert south-east of Russia, the Caspian basin. The principle is to make the fullest use of the land at a number of different levels of utilization, depending on the soil and position. Low-lying plains are being irrigated by gravity feed from reservoirs. Higher-level land is also being irrigated by water pumped up by electric power. Beyond the area of permanent irrigation range stock-rearing lands with piped water or electrically pumped wells. In the open desert sands are being fixed by saxaul trees, and even the sun is being turned into use for water-pumping and refrigeration.

The unit of planning is the whole river-basin. Great rivers, like the Volga, Don, and Dnieper, are being converted into a sequence of lakes, separated by dams with locks and power-stations, and sending out tentacles of irrigation canals. Floods and droughts will be balanced out. Already over one dozen major power stations are running and this trend has spread to the great Asian rivers. The new Bratsk power station in Siberia is the largest in the world. The Amu-Daria river is being extended and with its network of irrigation canals is gradually reclaiming the Kara Kum desert and will soon reach its original outlet in the Caspian Sea.

The power generated at the station will be used for industry and farming as well as for irrigation. These multiple uses will spread the load and increase the power factor so that the fullest use, both physically and chemically, will be made of every drop of water. Not only the great rivers but every little tributary is pressed into service. Each collective farm or small group of farms is encouraged and helped to set up its

299. The diversion of the Nile waters from their ancient course into a new permanent bed for the Aswan dam scheme. Here the waters enter the new bed.

own dam, reservoir, and power-station. Everywhere the emphasis is on mixed systems of cultivation combined with laboratory and field experiments. The object is at the same time to discover and to practise an agriculture which protects and enriches the soil with a well-balanced animal and plant ecology.[6.200]

It is the new large-scale civil engineering machines that have made possible the transformation of a barren into a flourishing countryside, compressing within a few years the improving farmers' work of centuries (p. 524). Giant trucks, bulldozers, drag-line and hydraulic excavators, do the work of a thousand times as many men. Already they are working on the scale of Nature. Geography can no longer be taken for granted, the world surface will henceforth be what man chooses to make it. For example, in the Baltic lands the relics of the Ice Age – moraines, erratic boulders, peat mosses produced by choked drainage – make the country poor and unproductive. What peasants working with their hands for millennia could not do machines are doing in a few years, clearing the stones and boulders from the land, cutting new rivers to speed drainage, getting power for all this from the peat, and leaving good farming land behind.

None of this is a Soviet monopoly. No sooner was the civil war ended than great conservation schemes were inaugurated in the People's

300. The Hsinanchiang hydro-electric power scheme, Chekiang province, China. Built by Chinese personnel.

Republic of China. The Huai river, the floods of which regularly devastated the richest eastern provinces, was tamed in a year, and already an impression has been made by the use of tributary control and relief basins on the Yangtze and Yellow Rivers. All this was at first done without waiting for more than a minimum of equipment, digging earth with hoes and carrying it away in baskets. This could have been done at any time during the last 6,000 years, but emperors and mandarins were impotent, for they could do nothing, even had they wanted to, without the active support of the people.

Since the early days, this heroic but slow method is being rapidly replaced by full mechanization. A great dam and powerful electric station has been built on the Yellow River, thus ending for ever the disastrous floods which have plagued northern China and helping to provide water for their productive but arid land in the dry seasons. None of this work is as yet sufficient to stop the effects of the terrible years 1959-62, but such droughts and floods, if they had not been held in check, would have in the past certainly caused famines involving tens of millions of deaths. By the time all the works are completed, which should not be more than a decade from now, the danger of drought and flood will have been permanently eradicated from the whole country.

There is scope enough in all the rest of the world for all the peoples, through the use of techniques and science, to transform their own countries. Some of them are indeed doing so. India has made a good start with the Damodar and other projects, but it is hampered by lack of capital.

The great Aswan Dam, over whose construction the world nearly went to war in 1956, is now nearing completion thanks to Soviet aid, and all over the under-developed parts of the world new dams are being projected or built. The first of the great west African rivers is being similarly controlled by the Volta Dam in Ghana. These works are an earnest that the whole face of the world could be changed for man's benefit.

All these things could be done by the peoples of each of these countries, if they were free from the direct and indirect effects of foreign domination. It could be done in less than a generation, if a fraction of the engineering potential now wasted in armaments were turned over to help men conquer Nature instead of destroying each other. The skill and money of the Americans, now wasted on atom bombs and super jets, would find a fruitful and exciting use in trying to beat the Russians at their own game – of changing Nature not for profit but for use.[6.161]

301. Mechanization of farming and the application of fertilizers by machine has transformed agriculture. Between-row cultivation and fertilizer spraying in soya bean field, Chiusan State Farm, China.

### THE TRANSFORMATION OF AGRICULTURE

The control of water, though essential, is only one part of the general transformation of agricultural methods which is rapidly occurring throughout the whole world. What has been done on impulse of the great urban food market of the industrial countries can now be spread to the newly liberated under-developed countries of the tropics and much improved in the process. This is an agricultural revolution which is far greater than that initiated in Britain in the eighteenth century. Basically it relies on fertilizer and biological control of plants and animals actually to grow the food, and on mechanization to ensure that only a minimum of human labour is required to plant, tend and harvest it. Its extent can be shown by the fact that in the United States one hundred years ago it required twenty men in the field to provide one man in the city with enough food, the rest they ate themselves. Now one man in the field can support twenty people in the city, with the aid of the fertilizers and machinery which the industry of the city men can provide.

It is evident that such a change requires in the first place a large capital equipment. To attempt to carry it out without capital means a simultaneous development of industry and agriculture which requires heroic efforts on the part of the people, as evident first in the Soviet Union and now in the People's Republic of China. Nevertheless, once the restriction of the old exploitative and feudal systems of agriculture are removed, this change is bound to take place and with increasing momentum. Now, in the light of the experience of the Soviet Union and of the People's Republic of China it should be possible to proceed even more rapidly by following the successes and avoiding the errors. Nor is it necessary to wait for the full development of machinery, although one of the great tragedies of the time is the fact that there is surplus capacity for agricultural machinery readily available in the capitalist industrial countries which could transform the whole world in a few years, whereas the autonomous self-help process is going to take anything up to four times longer.

As an example of what could be done immediately, we have the eight-point charter of agricultural production in China which lays down: (1) improvement of soil; (2) use of fertilizers; (3) extension and improvement of water conservancy; (4) improved seed; (5) improved planting methods; (6) better protection of plants; (7) better field management and (8) tools reform. All these measures are within the capabilities of the peasants of the local communes, and indeed would require communes to carry them out, especially in countries where water conservancy projects cannot be treated on a field or even farm basis.

The new agricultural revolution puts a premium on real agricultural science, on agrobiology, agrochemistry and agrophysics. Agricultural research has already doubled and quadrupled yields; it can only be effective when it is applied by people who have an interest in it and who also can acquire the necessary scientific knowledge. It is evident that in many fields the new advances in understanding of biology will have an enormous effect on agriculture. Developments of soil science and ecology should determine the best use of the land. The great power of fertilizers in multiplying yields should be combined with an improved fertilizer industry which can cheapen and improve them. Finally, the long battle with diseases of animals and plants has to be definitely won and the product of the land kept for human beings and not for insects. Through detailed biological studies it now seems likely, for instance,

302. The devastation caused by locusts in an orange grove, Sous Valley, in the Taroudant area of Morocco. The fruit is scattered on the ground, leaves and even bark have been stripped away. Spraying breeding locations and killing at the 'hopper' stage are the only effective remedies.

that the perennial plague of locusts can be finally eradicated from the world. The world cannot afford the time in which the peasant, however hard working and traditionally skilled, was illiterate and incapable of understanding science. It should be evident that the agricultural worker and the agricultural scientist, they may be both the same person, have to have a wider and deeper grasp of science than is required in any other branch of human activity. For the land to be fully used it will have to be entirely in the hands of the 'aristocrats', the 'slaves' will be machines.

THE POPULATION PROBLEM

The developments of the last twenty years, with the increasing area of liberated countries and the prospective end of direct colonialism, combined with the absolute need for more food to meet the growth of population makes an illiterate peasant agriculture impossible to maintain *in words* although it can still remain effective *in fact*, despite the efforts of United Nations organizations such as FAO, and voluntary campaigns against hunger. It is even difficult today to justify the acceptance of the old, natural checks of population of Malthus – hunger and pestilence – war, unfortunately, is still only too much with us. Nevertheless, there remains the enormous gap between what could be done and what is being done. With the population growing at 2·1 per cent per annum, a rate which seems most unlikely to diminish,[6.229] the idea of solving the problem by maintaining capitalism and preventing the poorer peoples from breeding seems to me doomed to failure.

This does not mean that conception and birth should not be scientifically controlled, but this should be in the interests of mothers and fathers of families, and not of those who wish to maintain their status as a comfortable *élite*, in a world kept, as Sir Charles Darwin (1887–1962) used to say, 'in the golden age of 1900'.

What has been done under the impetus of socialist ideas and practice already points to an enormous extension of civilization – agriculture and industry – together – in which the soil will not merely be preserved but indefinitely improved, and the life it supports will be multiplied. In the light of this knowledge and experience, all the talk of the danger of over-population appears all the more clearly as reactionary nonsense. This revival of Malthus in twentieth-century form is based itself on undeniable facts drawn from capitalist countries or their dependent empires. As such, it merely shows the fundamental failure of capitalism at the elementary job of keeping people alive. But – as the real operators of capitalism think but do not find it prudent to say – that was never their function. If it does not pay to keep people alive – well, then, let them die.

It should be clear by now that there is no possibility of raising the standard of living of the peasants of backward countries without a complete break with the old landlord or plantation system or even with a nominally free peasantry shackled to foreign companies. The fate of Malaya, the Philippines, and the banana republics of South America show this well enough.[6.171; 6.174] Real economic independence must be based on a growing industrialization, needed to use off-season labour and to provide the necessary equipment for a scientific agriculture. The policy started in the Soviet Union and now being followed in China and India, is based on the realization that the right way to secure adequate food production is by concentration of population in towns where they can produce, by intensive methods, the production goods, machines, and especially fertilizers for a full extension of the cultivated area on a basis of conservation agriculture on a high level, and at the same time the consumption goods for a far smaller agricultural population at a higher standard of living (p. 823). Any general adoption of back-to-the-land mysticism would result, even at the present level of population, in repeated famines.[6.191] It would be over-optimistic, however, to imagine such an overall scientific development of agriculture occurring in those parts of the world that are still in the orbit of capitalist economy. Foreign capital is not forthcoming, and domestic private capital is too small and too set on immediate profits. The only way, as India and Egypt are already finding out, is some form of socialism. Other under-developed countries will not be slow to follow their example, especially now that capitalism has lost its monopoly of technical know-how (pp. 830 f.). It is indeed the only possibility of getting out of the vicious circle of population growth to the limit of subsistence on a miserable standard of life.

The so-called 'population explosion' is not something to be deplored and halted but is rather a challenge to provide for the people who are to come and who are needed to build a new world. Of what is known and what has already been done to use science for the elementary task of providing people with food there is little trace in any neo-Malthusian book. Yet this represents only the merest beginning of what applied biology might do. The increase of world population is in itself not catastrophic; it is running at about 2 per cent per annum, and with a higher standard of living the rate is likely to be lower. The most that is required, therefore, for a rising food consumption is an average increase at a slightly higher rate. A 2 per cent per annum increase is well within the bounds achievable with present techniques. The application of new research will be essential only at later stages when there is a serious shortage of available land.

This is at present very far from being the case. The FAO estimate that of the 33 billion acres of the world's land just over 3 billion, or about 9 to 10 per cent, is cultivated.[6.178] Much of the remainder could be brought under cultivation, particularly in the equatorial areas, by a limited amount of real capital in the way that is already being done in the USSR and China. A conservative estimate by the geographer

L. D. Stamp indicates that some 10 billion people, or more than four times the present population of the world, could be maintained with present techniques at an adequate nutritional standard. That, with the present rate of increase, should see us well past the year 2100, and by that time people will be in a far better position than they are now to know how they want to solve the food and population problem. If they decide to go on increasing there will still be plenty of land for more scientifically directed exploitation, more especially in the desert belt, and the seas have only begun to be exploited. There is also a factor of between five and ten times to be picked up in more intense utilization of the land already cultivated. The present average yields are less than a third of the maximal, which are still very low; they could certainly be raised by biological research to far higher levels. Of the actual vegetable matter grown with such trouble, about four-fifths is burnt or ploughed in. It is by no means necessary that

303. Machine for extracting protein from fresh leaves at the Rothamsted Experimental Station, Harpenden. The crop goes up the elevator on the right of the photograph, and then into the pulper (centre). From this it flies out under a cowl and on to the far side of the circular table of the press. The table moves in step with the ram on the left; the ram is forced down and then lifted after 5–10 seconds by the cams shown on the left. Juice, containing protein, collects in the tray below the ram. In the photograph, waste from a pea-canning factory is being used. Grass may also be used, but it is one of the less suitable crops.

this should be so. The rich proteins produced in green grasses can, for instance, as Pirie has shown,[6.214] be extracted by pressing, and used for animal and at a pinch for human consumption, while the remaining cellulose is good cattle fodder. In this way, from the same meadow, a farmer could get bacon and eggs to supplement his beef, milk, and butter. Even greater possibilities are furnished by the use of yeasts and fungi to produce food from waste vegetable materials, or of algae for controlled photosynthesis. A promising result of how one of the immediate problems of malnutrition can be dealt with is that of the provision of proteins for topping up or completing the protein-poor diets of many tropical countries. It has been found by Champagnat that the bacteria can be grown economically and on a massive scale on raw petroleum, attacking only the paraffins and thus actually improving the quality of the oil. The protein extracted from the bacteria is of high quality and could be used not actually directly as a human food but as a condiment or as animal food. The conversion ratio between fat and protein is some ten thousand times greater than that of feeding the same amount to animals.

It is academic to dispute exactly how much food could be raised by scientific methods, for the methods themselves will grow and change in proportion as they are used. All this could be done by the use of conventional sources of energy. Now that nuclear-fission energy has been produced and even greater amounts from nuclear fusion may be reasonably expected (pp. 759 f.), the long-term prospects of food production are practically unlimited. As long as conventional agriculture is being used, atomic energy can provide the necessary water and heat (p. 851), but as the population rises to a thousand or more times its present figure, other or more direct methods can be invoked and will certainly be forthcoming, including atomic transmutation if necessary. The ghost of Malthus is well and truly laid.

None of this, however, can carry comfort for those who at this present moment have not enough to eat. The real difficulties are here not the scientific and technical ones. It is rather the achievement of the social and economic conditions that would make science applicable. If the grip of imperialism could once be shaken and the diversion of technical resources to war preparation were stopped there would be ample resources for the mechanical and chemical capital necessary to transform agriculture within a decade, together with ample funds for scientific research and development. The unused capacity of the United States motor industry would provide enough tractors to China in one year to increase grain production by 50 per cent. In 1951, it was estimated by a group of experts appointed by the Secretary-General of the United Nations that an annual investment of $19 billion would be sufficient to raise the standard of living of under-developed countries by 2 per cent per annum.[6.135] The factor for safe advance is about 6 per cent. However, as today something of the order of $100 billion is directly or indirectly spent on war preparations, such an increase is therefore immediately feasible.[6.19;* 6.134]

War, however, still remains the most profitable investment, and neo-Malthusians would be well advised to attend more seriously to that curse of humanity. If they could stop it, they would no longer have any need to invoke pestilence and famine to trim humanity to their genteel standards.

SOCIAL MEDICINE

The transformation of agriculture is only one aspect of the impact of modern biology on society; the other is the corresponding transformation of medicine. The great contributions to medicine of biology, and particularly biochemistry, of the twentieth century – vitamins, hormones, antibiotics, radiology, and radiotherapy – are only a part of a much greater transformation from the healing art to a science of health. Largely under the pressure of working-class protest, armed with the doctrines of an emergent socialism, disease began to be seen less as a punishment and warning from heaven, or even as the natural consequence of evil living, drink, and dirt, and more as a reflection on the conditions of life imposed by a heartless and stupid social system.

Social medicine, starting with the collection and analysis of medical statistics,[6.195] began to show in cold figures what had long been obvious, that the prime cause of illness was poverty.[6.197] Occupational diseases were the first target. Despite the obstinate obstruction of those whose profits seem to depend on the sacrifice of human lives, the most obvious of these – the lead poisoning of painters and pottery workers, the phossy jaw of match-makers, the silicosis of miners and steel-grinders – were denounced, and after many years some measure of protection and compensation imposed by law, though even today some 800 die annually in Britain of such diseases. The soot-laden air of the industrial cities still takes its toll; five times as many people die in Manchester of bronchial complaints as in the south of England. The preventable smog of London killed over 400 people in two days in 1952.

The greatest achievement of nineteenth-century social medicine was sanitation. It wiped out in industrialized countries the water-borne

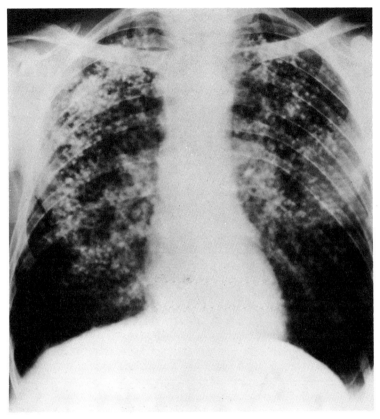

304. One of the occupational diseases of miners is silicosis, but now intensive research is being undertaken into preventive action. This X-ray photograph of the
lungs of a miner in an advanced state of silicosis (and from which he subsequently
died) shows the infection in the upper left of the picture as light patches.

diseases of cholera and typhoid, but the great killers, tuberculosis and
infantile diseases, still remained. They were to yield in the twentieth
century to better housing, better health measures, and most of all to
more and better food. The social value of the discovery of vitamins
lay not so much in the provision of the vitamins themselves as in the
interest it focused on nutrition as a prime requisite to health, particularly for children. Slowly but surely, despite the set-backs of slumps

and wars, the food of the favoured industrial countries has improved, and with it has gone a decrease in tuberculosis and a dramatic fall in infantile mortality.

The successes here have shown up in their full horror the unnecessary disease and death of less favoured countries. If one infant in fifty dies in Sweden, why should one in six die in India? It is now plainly evident that for lack of food or medical care two-thirds of the people of the world are dying avoidable deaths, and that out of ten infants that now die nine could be saved. To know that and do nothing about it is complicity in murder only less direct than acquiescing in killing them by atom bombs or napalm.

## NATIONAL HEALTH SERVICE

This knowledge has, however, not been without effect. In the last fifty years, all over the world, except in the citadels of individualism where health, like everything else, is for sale, has come an effective demand for free health service as a right. Even in Britain the medical profession has loyally, if unwillingly, acquiesced in a National Health Service. It is still a health service in name rather than in fact. Defence of national interests has seen to it that Britain has built few new hospitals and health centres since the end of the war. For the most part the National Health Service still depends on the old surgeries, where overworked doctors dispense ineffective drugs to queues of patients, and give them advice that they cannot take. Nevertheless it could be the beginning of a new attitude towards health, one in which the prime consideration was the right of every child, woman, and man to the biological and social environment that would best secure them a full, active, and healthy life. The doctor would still be needed, but rather as a counsellor and watchdog than as a patcher-up of bodies twisted and broken by bad conditions.

Social medicine logically implies social production and social distribution; how otherwise could everyone be guaranteed the work, rest, and food that are good for them? In brief, it implies socialism, and that is why in America particularly it is so fiercely resented, striking as it does at the sanctions of want and misery that lazy and greedy people think are the only means of setting the idle poor to work.

By contrast, wherever popular forces have triumphed there has been an instant drive for improved health services, especially for the children. By raising the status of doctors and nurses, by removing the need to compete for the few fee-paying patients, the old rooted objection of the medical profession to an increase of its members has been overcome.

For instance, in the territory which is now known as Uzbekistan, in tsarist times there was one doctor for every 31,000 of the population; in 1960 there was one for every 750, and Azerbaidzhan had one to every 450. These figures may be compared with Britain, where there is one for every 860, and Nigeria, where in 1960 there was one for every 33,000.[6.176; 6.228]

Even more spectacular has been the advance in China. There the drive for health has taken a mass popular form. The first stage has been the wiping out of sources of infection. China was one of the most fly-ridden countries of the world; after two years of popular government hardly a fly can be found in any Chinese town or village. The endemic centres of plague have been cleared up, and over four hundred million people vaccinated against smallpox. The medical services have been greatly increased. In north-east China, for example, by June, 1952, there were twenty times as many hospitals in factories and mines and twelve times as many health clinics as compared with pre-liberation days. A doctor is now available there for every 625 workers. Factories have been built to produce new life-saving drugs, defeating the cruel intention of the American ban on their importation.

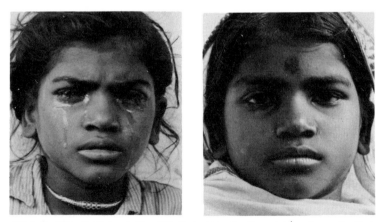

305 a, b. The effectiveness of antibiotic treatment often appears dramatic. Two photographs by Homer Page for the World Health Organization showing a child before and after treatment with aureomycin for an eye disease that could have resulted in permanent blindness.

A similar transformation could be achieved in all the unhealthy tropical and sub-tropical areas, unhealthy only because poor and exploited. It can be done by the people themselves, and only by them. Medical help from outside, however well intentioned, can only be palliative, and sometimes not even that, when, in the absence of land reform, it only leads to general impoverishment. In the Bengal malaria epidemic of 1944 the free issue of drugs found its way rapidly into the black market, the recipients preferring the risk of death from disease to the certainty of it from starvation.

In the last fifty years the science of biology and the practice of social medicine have proved that man is already capable of lifting the burden of disease and death that has weighed him down for millennia. Now that that is known, nothing, not even the greatest development of hydrogen bombs and super poisons, is going to prevent the mass of humanity from finding a full and healthy life.

### 11.10 The Future of Biology

This account of the present situation of biology and its effects on society should bring out by its very length and diffuseness the increasing number of ways in which the newly won knowledge is impinging on the lives of almost every person in the world. Growing, as it has, hardly less fast than have the physical sciences, it has had a far more rapid impact on society than they have had – except in the service of war. A new drug or a new breed of plant can be put into service much more quickly than a new method of building or engineering, or even a new aeroplane. The turnover of biological science is quicker, the capital cost smaller.

This consideration looked at in another way means that biology is not so directly linked to heavy industry. This is a major reason why the financial support and the number of research workers in the biological are so much less than in the physical sciences.

In the near future, given an end to the Cold War, the rewards of biology are bound to lead to its very rapid increase. At the same time its intrinsic interest will draw more and more able workers into biological studies. Apart from nuclear physics it is biology, and particularly biochemistry and biophysics, that are already the most exciting fields of research. This is because biology offers problems of great complexity where ingenuity is at a premium. From what has been found in the last half-century it is now apparent what an altogether too limited and simplified view of organisms and their interactions was held by earlier investigators. Even the simplest of them surpasses a thousand or a millionfold in absolute complexity the most complicated systems devised by man. Indeed, if the early biologists had formed any idea of the order of complexity of the subjects they were tackling they would probably have lacked the courage to undertake the task, for, as Marx pointed out, man does not attack problems unless he already has the means of solving them. As the army of biological science increases, so also will the complexity of the problems which it will attack.

## THE BREAK-THROUGH IN BIOLOGY

It should already be evident from this account that a major break-through in biology in the understanding of the mechanisms of nucleic-acid–protein synthesis reported on pp. 915 f. has in fact occurred in the last five years, but it will be long before its effects are fully felt throughout the whole field and particularly in its applications to medicine and agriculture. It is a break-through not only in the field of theory, but even more in that of experimental techniques. The new chemical and physical methods are already transforming our concepts not only of biological structures but also of biological functions. Electron microscopes, tracers, and electronic detecting devices between them are providing us with new dimensions in biology. The new concepts of polymer chemistry and statistical techniques, aided by electronic computers, can help us to analyse their findings, ranging from the interior of viruses to the behaviour of animal societies. For the immediate future we can already foresee the follow-up of the great break-through, an extension of the new methods into more and more fields and the final mopping up of whole areas of ignorance. This process is certain to reveal further areas of ignorance but a great deal of country will be won and can be used. Biology is clearly becoming an intelligible, logical subject and has moved far from the merely natural-history approach of former ages. We need to appreciate to the full the revelations on the structure of the cell and its biochemical significance, the fuller understanding of the role of the nucleic acids in their various forms in the genesis of proteins and particularly the almost untouched field of lipoid structure. These clearly have a large part of play in the control of processes occurring in cells as well as in multicellular animals. At the other end of the scale the prospects of unravelling the problems of the internal nervous control of organisms and the communication between them also seem to promise great advances. Linked with both of these are the prospects of creating a more self-consistent picture of heredity and evolution, stretching all the way back to the origin of life on this earth and forward to the emergence of human societies.

## TOWARDS A NEW BIOLOGICAL THEORY

The prospects of multiple advances over an enormous field emphasize the ever-greater need of co-operation. Effective biological advance is necessarily a vast combined operation regardless of whether it is recognized to be, for the worth of each man's work depends on that of dozens of others. It calls for a well-directed information service and some sense of strategy (pp. 1271 f.) which will not prevent the recognition and exploitation of the unexpected.

Biology cannot in the nature of things be as simple as physics or even chemistry, since it includes these subjects in itself. Nor can it be expressed in the language of precise mathematics, because it has too great a multiplicity to describe by enumeration. Indeed, most attempts to reduce biology to mathematics, because of their very abstraction and inappropriateness, lead to errors which would not have been made had the same ideas been expressed in words. Nevertheless, because we must be able to deal with living systems practically, an appropriate language has to be found to describe them, think about them, and thus control them in a rational rather than a traditional way. Any useful biological language must treat of the structure and behaviour of organisms as such and be adequate to their higher level of complexity. Even through the chaos of present-day biological discovery and controversy we are beginning to see the form that language will take.

## NEW GENERALIZATIONS

The great new discoveries in molecular biology are already leading to new generalizations about the nature of life. Life can now be defined in much more precise terms and at the same time the conditions in which it can occur can be more clearly seen. Life could be described not in the more limited form given by Engels as 'The mode of motion of protein' but as the mode of production and reproduction of identical molecules. What we see on the outside of life in organisms is really a reflection of the structures inside the molecules. Molecules reproduce and multiply before organisms can. It is now becoming more and more evident that new generalizations of great importance in biology are about to emerge. The central discoveries of biochemistry, pointing to the underlying chemical nature and chemical origin of life, have yet to be translated into a general biological theory. Such a theory must be intrinsically evolutionary; that is, it must bring out the character of the present as a resultant of the past embodied in biological structures and functions. The older approach to evolution was based on visible appearance and performance, the new must look right down to the scale of atoms, though in doing so it must never lose sight of the larger unities of organisms and societies. Just because it has to involve matter and history together, it can only be built along the lines of dialectical materialism. The mechanical theories derived from the Newtonian period cannot cope with the essentially historic aspect of biology. In physics it is generally sufficient to know how systems work. In biology it is equally important to know how they got to be that way. The whole drama of evolution is an example of a serial production of new forms arising essentially out of the conflicts engendered inside the previous stages, owing to inevitable contradictions within the organism and in its relation to its environment.

### THE ORIGIN OF LIFE

The questions of the origins of life were discussed in earlier editions in the biology section itself. The rapid changes in our appreciation of the problems of the origins of life and the new evidence that is coming in rapidly from the study of objects from space have so altered the picture, and at the same time made it one of the most exciting in science, that it has become impossible to arrive at any kind of fixed conclusions. It should therefore be treated as part of the future rather than the present in biology.

For a long time the question of the origin of life was put outside the pale of regular biological discussion, being given a certain theological flavour, associated with the problem of spontaneous generation, which was taken to be a hangover from mystical and religious ideas of the past. In fact, in the past the question of spontaneous generation did not raise any particular metaphysical concepts, because it was considered to be quite natural, owing to the false observation of nature – flies being generated in meat and frogs in mud by the force of nature, which could, in any case, do wonderful things. People preferred to believe that barnacles gave rise to geese and that swallows spent the winter under the mud of ponds.

It was only in the nineteenth century that more rigid scientific disciplines were adopted and the classical work of Pasteur (p. 649) showed that, in the normal conditions of the laboratory, spontaneous generation did not take place. This was interpreted, particularly by vitalist

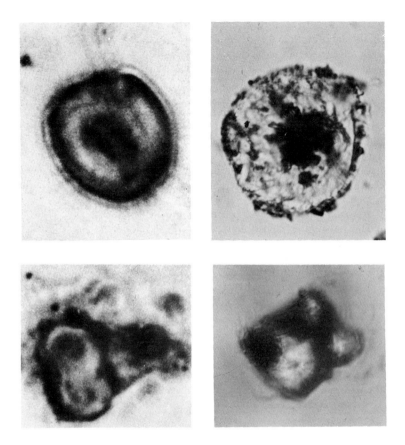

306. Comparison of microscopic organic material of terrestrial origin with shapes found in carbonaceous meteorites. The photograph (top, left) shows an element of material from a meteorite found at Orgueil (after *G. Claus* and *B. Nagy, Fordham University, New York*) and (top, right) cored olivine globule from Alais meteorite (after *G. Mueller, Birkbeck College, London* and published originally in *Nature*, 1962, vol. 196, pp. 929–32). Below (bottom, left) element from Orgueil meteorite (after *Claus and Nagy*) and (bottom, right) short ragweed (*Ambrosia elatior*) pollen grain. (After *F. W. Fitch and E. Anders, Enrico Fermi Institute for Nuclear Studies, University of Chicago*). These photographs show (top) evidence for water-containing materials and (below) evidence for organic remains. The combination of shapes and water-containing minerals is an indication that life may exist elsewhere in space, but the evidence is not yet conclusive.

philosophers, to show that there must be something not reproducible in the laboratories, some 'life-force' which generated organisms. In view of Darwinian evolution, however, all that it was necessary to generate was the primitive organism: the others could be evolved from it.

It was only in the twenties of this century that Oparin and Haldane put forward the hypothesis that life could have been generated biochemically on the surface of an earth which possessed, at the beginning, a different atmosphere – reducing rather than oxidizing. Thus sunlight could build up organic compounds starting from very simple hydrides like methane, ammonia and water and these could later develop through an intermediate colloid or coacervate stage into organisms.[6.177; 6.212; 6.220]

At that time the development of biochemistry was such that it would have been impossible to trace this development in any greater detail. But in the forties and fifties of the century, and particularly now, both the possibilities and the difficulties of the generation of life have been more appreciated.

The great break-through of the nucleic-acid–protein synthesis of reproduction is, from the point of view of life, what might be called the 'end point' of the genesis of life. From then on we are in the normal field of biochemistry and evolution. Recent changes have been in the ideas of the first stages – the building of the elementary small molecules, amino acids, purines and pyrimidines, that go to make up the metabolic units of the life process. It is already evident that these can be formed in the absence of life. This was first done in the laboratory in 1953 by Urey and Miller. Meanwhile all the evidence for its having been done in space already existed in museums in the form of the rare carbonaceous meteorites which are shown to contain, among others, these kinds of compounds. They are to be found in a fairly consolidated form, not only in the meteorites, which may be parts of some earlier planet or asteroid, but also in the actual, probably primitive, cosmic dust that rains down all the time in the upper atmosphere and gives rise to such phenomena as the noctilucent clouds seen in high latitudes between the auroral level and that of the shooting stars at about eighty kilometres above the surface of the earth.

These particles seem to be composed very largely of nickel containing iron and covered with a thin, presumably organic, skin, itself probably produced by the interaction of cosmic rays and the simpler and more volatile methane and ammonia compounds. If this is true, and it requires a great deal of confirmation, complex carbonaceous substances are in the first place formed in outer space and must be extremely common. The material formed in this way will be built into the planets in their first formation and may be gradually sweated out of them and chemically altered in the process as they are consolidated. There is plenty of carbonaceous material in the igneous rocks of the earth and some is well known to us in the form of diamonds, which incidentally are now shown to contain nitrogen as well as carbon.

However this may be, we have now an entirely independent source of the primitive soup from which organisms can arise. This soup contains not only the necessary elements but also the energy-rich compounds which will provide the driving force for further chemical change. This makes the origin of life a much more probable process than was considered previously. We have not so much the problem of *how* life could have originated, because there are so many different ways it could have done, but to find precisely the method by which it actually did originate. It is likely that further progress in this field will be extremely rapid. Claims have been made for the detection of actual organisms in meteorites, but these seem now to have been discounted. The present studies link the question of the origin of life with that of the origin of the sun and the solar system.

I will touch on the other now conceivable study of life, that on other planets. Armed with the new biochemical knowledge, we may come to be in a position to study not biology, which is really terrestrial biology, but comparative biology. Just as the biologists of the nineteenth century were able to work out the theory of evolution by looking at the different ways in which the higher organisms and plants actually managed their existence, so will we be able to find out how life originated on this planet by comparing it with life on other planets. Philosophically, the important aspect is the linking over a whole field of the concept of origin, structure and function, an idea already adumbrated by Marx in relation to societies and by Engels in relation to aspects of natural science.

APPLICATIONS OF BIOLOGICAL RESEARCH

These considerations, though academic in appearance, are none the less potentially of the greatest social and economic importance. The nearer we approach to a fundamental understanding of biology, the sooner we can acquire conscious control of our living environment and our own bodies. Already many existing prospects are opening up.

Greater knowledge of soil science, ecology, and plant physiology should provide far greater crops and make their yield reliable. The possibility of controlled heredity should lead to the creation not only of improved but also of new types of food plants adapted to all climates. There is an infinite range of possibilities of producing new foods and drugs through the cultivation of yeast, fungi, and algae. An intelligently applied biochemistry should see that we get the fullest and best use of the foodstuffs we produce and make cooking a science while preserving its triumphs as an art.

Advances in medicine touch us even more directly. Deeper knowledge of biochemistry and physiology should bring us closer to the adjustment of bodily functions to environment so that disease is detected as a tendency to avert rather than as a condition to cure. The knowledge, which we do not yet possess, of the optimal diet over the whole life cycle, adapted to each individual, should reduce to a minimum the need for drugs. At the same time drugs designed for the purpose would be on hand to check the unprevented failure of any part of the metabolism. With infective disease and cancers rendered harmless, if not abolished, interest will centre on the prolongation of life and health to well beyond the present limits. Surgery can already repair and temporarily replace such vital organs as hearts and kidneys. Deeper knowledge of cell metabolism should enable us to regenerate damaged organs. Mental diseases, in so far as they are of a physiological rather than social origin, should yield to advances in the knowledge of neural functions.

All this and far more as yet unimagined will come with a speed more than proportional to the research effort now expended in the field of biology. This could easily and rapidly be increased by a transfer of research effort from the military field. The problems of an increasing population and of static or decreasing food supplies can be solved only by an active and advancing biology. No social system, however respectably established, can resist indefinitely such urgent human demands, especially if at the same time there are practical demonstrations of the full use of biology in the service of man. The future of biology is itself as much a social as a biological question, and the changes which the forms of human society are likely to undergo in this period of transition are certain to transform the science of biology as well as the biological environment of mankind.

Table 7

**Biology in the Twentieth Century (Chapter 11)**

The biological sciences in the twentieth century are shown in this table (next page) with the same time division as the physical sciences. The columns are arranged to correspond approximately to the sections of Chapter 11, as indicated. Only a few of the major advances are shown. In biology even more than in physics a single research may be continued for periods of twenty years – for instance, Landsteiner's classic work on blood groups covers the first three decades of the century. The dates of the entries marked in the table are accordingly somewhat arbitrary. An attempt has been made, however, to indicate the date of the most decisive results of the research.

| | Historical Events | Biochemistry | Microbiology | Medicine |
|---|---|---|---|---|
| 1890 | Colonial wars | Rise of biochemistry | | |
| | Growth of monopolies | *Buchner* enzymes | *Ivanowski* plant viruses | *Eykmann* beri beri Nutritional studies |
| | | | *Löffler* animal viruses | |
| 1900 | Russo-Japanese War | | *Landsteiner* blood groups | *Hopkins* VITAMINS |
| | First Russian revolution | *Wilstatter* photosynthesis | | |
| 1910 | Increasing inter-imperialist tension | | | *Ehrlich* chemotherapy salvarsan |
| | | *Henderson* 'Fitness of the environment' | | |
| | First World War Russian Revolution | *Warburg* respiration enzymes | *Herelle* bacteriophage | |
| 1920 | | | | HORMONES *Banting* insulin *Doisy* ovarian hormone |
| | Post-war depression Fascism in Italy | *Svedberg* ultracentrifuge | | |
| | General Strike in Britain | | | *Minot* pernicious anaemic factor |
| 1930 | | *Sumner* crystalline enzymes | *Stanley–Bawden–Pirie* crystalline viruses | *Windaus* Vitamin D *St Gyorgy* Vitamin C |
| | Great Depression | | | ANTIBIOTICS |
| | Rise of Nazism Spanish Civil War | *Keilin* cytochrome *Krebs* biochemical cycles | *Engelhart* muscle as an enzyme | *Domagk* sulphonamides *Fleming, Florey, Chain* penicillin |
| | Second World War | | | |
| 1940 | Invasion of Soviet Union | *Perutz* X-ray studies of crystalline proteins *Martin, Synge* paper | *Avery* modified strain of pneumococcus Electron microscope | Large-scale manufacture of penicillin Other antibiotics |
| | Liberation Cold War | chromatography *Sanger* amino acid order in proteins | studies of virus and bacteriophage | Cortisone |
| | People's Republic of China | | | B.12 |
| 1950 | Korean War | | | |
| 1955 | Suez Hungary | Synthetic ACTH *Calvin* elucidation of photosynthesis | *Fraenkel-Conrat* virus infection by nucleic acid | *Gudov, Androsov* mechanized surgery |
| 1960 | Liberation of Africa Congo Cuba | *Kendrew* protein structure | Structure of viruses | Sera for poliomyelitis |
| 1965 | | | | |

| Cytology and Embryology | Control Mechanism | Heredity, Evolution, and Ecology |
|---|---|---|
| *Roux* and *Driesch* experimental embryology | *Starling* electrocardiograph | *Bateson* rediscovery of Mendel |
| *Loeb* artificial fertilization of eggs | **Pavlov conditioned reflexes** | *De Vries* mutation<br>*Schimper* plant ecology |
| | *J. S. Haldane* respiration | *Glinka* pedology<br>*Johannsen* pure lines |
| | *Sherrington* nervous systems | *Bateson* linkage |
| Many studies in fertilization and cell division | | *Morgan* genetics of drosophila |
| *Harrison, Fell* tissue culture<br>Development of organs | *Watson, Köhler* animal psychology<br>*Von Frisch* communication bees | Chromosomes and genes |
| | *Berger* electro-encephalograph | *Volterra* food chains<br>*Müller* X-ray induced mutation |
| *Spemann* induced embryo development, the organizer | *Adrian* electric nature of nerve impulse | *Fisher* statistical theory of evolution<br>*J. B. S. Haldane, Ford* ecology and evolution |
| | | *Lysenko* vernalization |
| *Ruzcka* and *Ardenne* electron microscope | *Young* behaviour and neurology of octopus | Genetics controversy in Soviet Union |
| *Wyckoff* study of organisms and tissues | | |
| | *Hodgkin* chemical changes in nerve | |
| Electron microscope resolution of intra-cellular structure<br>*Huxley* muscle fibre structure | *Grey Walter* analysis of brain currents | MOLECULAR BIOLOGY<br>*Watson, Crick* structure of nucleic acid, genetic code |
| Electron microscopic study of cell organelles, mitochondria, ribosome | | |

# Notes

For explanation, see page 699.

PAGE 703. *Some of my critics have doubted the justice of speaking of a second scientific revolution in the twentieth century, alleging that in this case there was no break in continuity of research such as occurred between Classical times and the Renaissance, nor was there any notable slackening of the pace of advance. Now the terms revolution and continuity are inevitably relative. I have even been attacked by other critics for under-rating the continuity of Renaissance and Medieval thought. But I feel that if we grant the term revolution in one case, we must grant it in the other. Against the revolution of the earth and the circulation of the blood, the telescope and the vacuum pump, and the upsetting of previous ideas that these implied, we may urge the discovery of the nuclear atom, relativity, and the quantum theory, as well as of the processes of biochemistry and the inner structure of the cell, the electron microscope, and the electronic computing machine. Add to that the sudden acceleration of all scientific activity and its application, from atom fission and television to the control of disease, and it would appear that if this is not a scientific revolution nothing is. Nevertheless, the contention that the two revolutions are not comparable may be true in another sense. The first revolution actually discovered the method of science, the second only applied it. The new revolutionary character of the twentieth century cannot be confined to science; it resides even more in the fact that only in our time has science come to dominate industry and agriculture. The revolution might perhaps more justly be called the first scientific-technical revolution (p. 712).

PAGE 704. *Dr Richter objects to my statement on the coincidence of the revolutions in science and in society. What I have effectively said here is that two revolutions occurred at about the same time. Actually the scientific revolution had a twenty years' start on the political one. However, I have no wish to imply that the scientific revolution actually caused the political one. As I explain, the theoretical ideas on which the scientific revolution was based were laboriously evolved after the first experimental breakthroughs. The study of gas discharges and the properties of the electrons led to the revision of classical physical theory. Hence, quantum and relativity theory were not originally produced by the political atmosphere, but were internally conditioned, helped on as I have indicated by relatively minor industrial applications. One direct connexion, however, was of vital importance, namely, Lenin's interest in science and its application, as evident, theoretically, in his *Materialism and Empiro-Criticism*, and, practically, in the determination from the start that the Russian Revolution should help and use science immediately –

'Communism is Soviet power plus the electrification of the whole country'. Successes in the Soviet Union, which were seen to be based on science, led reciprocally to a much greater interest in the use of science in the capitalist states. Especially since 1957, the existence of a new scientific–technological revolution has come to be recognized as much in capitalist as in socialist states.

PAGE 710. *Besides the geological departments of the older universities, three major institutes with a four-year course in geology have been set up. One of these, with several thousand students, occupies the palace built by the Japanese at Changchun for the former Emperor of Manchukuo. There are in addition separate institutes for mining and oil technology. Students from these institutes are already in the field, especially in the unexplored western parts of the country, where large new deposits of coal, iron ore, and minerals have been found.[6.17]

PAGE 715. *The arrival of a new technical revolution that was basically also scientific had long been in accordance with Marxist theory. It was recognized in the West, and particularly in America, only as the result of the first sputnik – a scientific and mechanical triumph of which the self-satisfied rulers of the capitalist countries had long believed the Soviet Union was incapable. Bertrand Russell had even gone so far as to say that 'an atom bomb built on Marxist principles would not work', unfortunately for him, only a month before the first Soviet bomb was exploded. The result of this appreciation of the need for science has been the formal acceptance of a rather frenzied, but not as yet effective, crash programme of increasing scientific education.

PAGE 740. *I am accused by Dr Richter of undervaluing the role of theory in science, particularly in relation to Planck's quantum theory. My account of the rise of modern physics is frankly somewhat biased on the experimental side, on account of my greater knowledge and experience of it. It is relatively easy to see now that many of the advances in chemistry in the nineteenth century, particularly physical and thermochemistry, are explainable only in quantum terms, but this was not seen at the time. The real triumphs of the quantum theory were in the explanation of the atomic phenomena which followed the completely unexpected and untheoretical discoveries of Röntgen and Becquerel. The Rutherford–Bohr atom was the first great vindication of the quantum theory as well as of the planetary model of the nuclear atom.

PAGE 744. *The understanding of the principles of the behaviour of matter at high velocities proved to be practically essential in the design of the great synchrotrons and cyclotrons of nuclear physics. More recently it has been thoroughly vindicated in the discovery of the super stars of millions of suns' masses which are found in the nuclei of some galaxies (pp. 771 f.).

PAGE 746. *Professor Rosenfeld, in a critical review, claims I have not given credit to Mach for his genuine physical insight but have damned him quite simply for his positivist philosophy.[6.119] If I had been writing on his mechanics and hydrodynamics, most relevant in this age of supersonic flight, I would have given him full praise. Here, however, I am dealing with the effects of his essentially subjective and sensational approach to physical theory – for example, his anti-atomism – which I still feel has done much damage and will do more to physics.

PAGE 816. *Major contributions in the space age have so far been in the field of exploration by non-human projectiles, bringing with them the steady and, indeed, very rapid improvement of methods of observation and retransmission of information. The first triumph, in which the Soviet Union sent round the moon a satellite which retransmitted photographs of the side which had hitherto never been seen, was followed, after a number of unsuccessful trials, with the magnificent success of the United States' Ranger 7 on 31 July 1964. The close-ups of the moon received from the Ranger, if they have not revolutionized our knowledge of the surface of the planet, at least showed that its surface was very much the same plain of volcanic or meteoritic origin, completely peppered by small fragments and finished off with a dust layer not of such thickness as would make a landing difficult. These results when fully analysed are likely to furnish much information, not only about the moon but also general planetary information and particularly about the structure of the earth.

Whether the enormous expense in planning the operation is justified is still a matter of question. But such questions will be answered within a few years, and long before any landing is attempted, by other indirect scientific methods of exploration which may give us the very necessary supplementary information about the chemical and physical nature of the moon's surface. There can be no doubt that the combination of national prestige and ancillary useful military strategy has enabled these explorations to be carried out, but in the long run we may be thankful because it sets the scale for expenditure in science which may be fruitfully followed over the whole field in happier times.

PAGE 942. *These methods offer, as already hinted at (pp. 855 f.), an ultimate possibility of direct communication. Grey Walter has shown that the brain current generated by one stimulus can be used to let off another. From there it is not far from learning how to produce an external signal by thought alone and then, through some code, effecting actual communication. One might hope in such a method to speed up the output of human thought beyond that of speech, at least to the speed of reading.

PAGE 977. *The 1962 UN *Report on the Economic and Social Consequences of Disarmament*[6.134] quotes estimates of the total amount of foreign capital required annually by the under-developed areas ranging from $6 billion to $10 billion for target rates of growth of per capita real income of about 2 per cent.

I have suggested in my book *World Without War*[6.19] (1961) that, with world military expenditure reduced to one third, there would be $20·5 billion in interest-free credits for the under-developed countries. I would now (1964) estimate the figure to be nearer $25 billion.

# Bibliography to Volume 3

## PART 6

1. ALLEN, J. S., *Atomic Imperialism*, New York, 1952
2. APPLETON, Sir E., 'Science for its Own Sake', *The Advancement of Science*, vol. 10, 1953
3. ARMITAGE, A., *A Century of Astronomy*, London, 1950
4. ASHBY, Sir E., *Technology and the Academics: an Essay on Universities and the Scientific Revolution*, London, 1958
5. AYER, A. J., *The Foundations of Empirical Knowledge*, London, 1947
6. AYER, A. J., *Language, Truth and Logic*, 2nd ed., London, 1947
7. BARAN, P. A., *The Political Economy of Growth*, London, 1958
8. BARBER, B., *Science and the Social Order*, London, 1953
9. BAUER, E., *L'Électromagnétisme hier et aujourd'hui*, Paris, 1949
10. BERNAL, J. D., 'The Answer to the Hydrogen Bomb', *Labour Monthly*, vol. 35, 1953
11. BERNAL, J. D., *The Freedom of Necessity*, London, 1949
12. BERNAL, J. D., *Marx and Science*, London, 1952
13. BERNAL, J. D., *Science for a Developing World*, London, 1962
14. BERNAL, J. D., *Science and Industry in the Nineteenth Century*, London, 1953
15. BERNAL, J. D., 'Science in the Service of Society', *Marxist Quarterly*, vol. 1, 1954
16. BERNAL, J. D., and CORNFORTH, M., *Science for Peace and Socialism*, London, 1949
17. BERNAL, J. D., 'Science and Technology in China', *Universities Quarterly*, vol. 11, 1956
18. BERNAL, J. D., *The World, the Flesh and the Devil*, London, 1929
19. BERNAL, J. D., *World Without War*, 2nd ed., London, 1961
20. BICHOWSKY, F. R., *Industrial Research*, New York, 1942
21. BIRKS, J. B. (ed.), *Rutherford at Manchester*, London, 1963
22. BJERKNES, J., *Investigations of Selected European Cyclones by Means of Serial Ascents*, Oslo, 1935
23. BLACKETT, P. M. S., *Atomic Weapons and East–West Relations*, Cambridge, 1956
24. BLACKETT, P. M. S., *Military and Political Consequences of Atomic Energy*, London, 1948
25. BLACKETT, P. M. S., *Studies of War, Nuclear and Conventional*, Edinburgh, 1962
26. BONDI, H., *Cosmology*, Cambridge, 1952
27. BORN, M., *The Natural Philosophy of Cause and Change*, Oxford, 1949
28. BOWEN, E. G., 'An Unorthodox View of the Weather', *Nature*, vol. 177, 1956

29. BRENNAN, D. G. (ed.), *Arms Control and Disarmament*, London, 1961
30. BRIDGMAN, P. W., *The Logic of Modern Physics*, New York, 1927
31. BRODIE, B., *Strategy in the Missile Age*, Oxford, 1959
32. BROGLIE, L. DE, *The Revolution in Physics*, New York, 1953
33. BROGLIE, L. DE, *Savants et découvertes*, Paris, 1951
34. BRUNSCHVIEG, L., *L'Expérience humaine et la causalité physique*, Paris, 1922
35. *Bulletin of the Atomic Scientists*, vol. 12, 1956, p. 270
36. BURHOP, E. H. S., *The Challenge of Atomic Energy*, London, 1951
37. BURHOP, E. H. S., 'The Origins of the Pugwash Movement', *Scientific World*, 1961, no. 3
38. BUSH, V., *Modern Arms and Free Men*, London, 1950
39. CARDWELL, D. S. L., *The Organisation of Science in England*, London, 1957
40. CARTER, C. F., and WILLIAMS, B. R., *Investment in Innovation*, London, 1958
41. CARTER, C. F., and WILLIAMS, B. R., *Science In Industry*, London, 1959
42. CAUDWELL, C., *The Crisis in Physics*, London, 1939
43. CORNFORTH, M., *In Defence of Philosophy*, London, 1950
44. CORNFORTH, M., *Science Versus Idealism*, London, 1946
45. COSSLETT, V. E. (ed.), *The Relations Between Scientific Research in the Universities and Industrial Research*, London, 1955
46. COUZENS, E. G., and YARSLEY, V. E., *Plastics in the Service of Man*, Penguin Books, 1956
47. CROWTHER, J. G., *Science in Liberated Europe*, London, 1949
48. CROWTHER, J. G., and WHIDDINGTON, R., *Science at War*, HMSO, London, 1947
49. CUSHMAN, R. E., 'The Repercussions of Foreign Affairs on the American Tradition of Civil Liberty', *Amer. Phil. Soc. Proc.*, vol. 92, 1948
50. DARWIN, C. G., *The Next Million Years*, London, 1952
51. DAVY, M. J. B., *Interpretative History of Flight*, HMSO, London, 1946
52. DEMBOWSKI, J., *Science in New Poland*, London, 1952
53. DENNIS, N., et al., *Coal is Our Life*, London, 1956
54. DE WITT, N., *Education and Professional Employment in the U.S.S.R.*, Washington, 1961
55. DIEBOLD, J., *Automation*, New York, 1952
56. DINGLE, H. (ed.), *A Century of Science*, London, 1951
57. DOBB, M. H., *Economic Growth and Underdeveloped Countries*, London, 1963
58. DOBB, M. H., *Essay on Economic Growth*, London, 1960
59. DUNSHEATH, P., *A Century of Technology*, London, 1951
60. EATON, J., *Socialism in the Nuclear Age*, London, 1961
61. EINSTEIN, A., and INFELD, L., *The Evolution of Physics*, Cambridge, 1938
62. EINZIG, P., *The Economic Consequences of Automation*, London, 1956
63. DUHEM, P., *Le Système du monde*, 5 vols., Paris, 1913–17
64. EVANS, I. B. N., *Rutherford of Nelson*, Penguin Books, 1939
65. FEDERATION OF BRITISH INDUSTRIES, *Industrial Research in Manufacturing Industry, 1959–60*, London, 1961
66. FEDERATION OF BRITISH INDUSTRIES, *Research and Development in British Industry*, London, 1952
67. FEDERATION OF BRITISH INDUSTRIES, *Scientific and Technical Research in British Industry*, London, 1947

68. FINDLAY, A., *A Hundred Years of Chemistry*, 2nd ed., London, 1948
69. FLEMING, Sir J. A., *Fifty Years of Electricity*, London, 1921
70. FLEMING, Sir J. A., *The Thermionic Valve*, 2nd ed., London, 1924
71. FORD, H., *My Life and Work*, New York, 1926
72. FREEDMAN, P., *The Principles of Scientific Research*, London, 1949
73. GELLHORN, W., *Security, Loyalty and Science*, Ithaca, New York, 1950
74. GIEDION, S., *Mechanization Takes Command*, Oxford, 1948
75. GLASS, B., 'Academic Freedom and Tenure in the Quest for National Security', *Bulletin of the Atomic Scientists*, vol. 12, 1956
76. GOLDSTEIN, W., and MILLER, S. M., *Theories of Terror: the Indelicate Premises of Nuclear Deterrence*, Woking, Surrey, 1962
77. GOODEVE, Sir C., 'Using Science to Reach Decisions', *The Manager*, May, 1953
78. GOUDSMIT, S. A., *ALSOS: The Failure in German Science*, London, 1947
79. HASLETT, A. W. (ed.), *Industrial Research in Britain*, 4th ed., London, 1962
80. HEATH, A. E. (ed.), *Scientific Thought in the Twentieth Century*, London, 1951
81. HEISENBERG, W., *Physics and Philosophy: the Revolution in Modern Science*, London, 1959
82. HERSEY, J., *Hiroshima*, Penguin Books, 1946
83. HMSO, *Committee of Enquiry into the Organization of Civil Science*, London, 1963
84. HMSO, *Government Scientific Organization in the Civilian Field*, London, 1954
85. HMSO, *Notes on Science in USA, 1954*, London, 1955
86. HMSO, *Statistical Summary of Mineral Industry*, Colonial Geological Surveys, Mineral Resources Division, London, 1954
87. HMSO, *United Kingdom Atomic Energy Authority: Second Annual Report, 1955-56*, London, 1956
88. HOYLE, F., *The Nature of the Universe*, Oxford, 1950, and Penguin Books, 1963
88a. JAMES, W., *The Moral Equivalent of War*, New York, 1910
89. JAY, K. E. B., *Britain's Atomic Factories*, HMSO, London, 1954
90. KAHN, H., *Thinking About the Unthinkable*, London, 1963
91. KURCHATOV, I. V., 'On the Possibility of Producing Thermonuclear Reactions in a Gas Discharge', *Discovery*, vol. 17, 1956
92. LANGE, O., *Disarmament, Economic Growth and International Co-operation*, Leeds, 1963
93. LAPP, R. E., *Kill and Overkill: the Strategy of Annihilation*, New York, 1962
94. LARSEN, E., *The Cavendish Laboratory*, London, 1962
95. LILLEY, S., *Automation and Social Progress*, London, 1957
96. MACMILLAN, R. H., *Automation: Friend or Foe?*, Cambridge, 1956
97. MARTIN, C. N., *The Atom: Friend or Foe?*, London, 1962
98. *Marxist Quarterly*, vol. 3, 1956, no. 2: articles on automation and atomic energy; no. 3: articles on the Twentieth Congress of the Communist Party of the Soviet Union and the Sixth Five-Year Plan
99. MELVILLE, Sir H., *The Department of Scientific and Industrial Research*, London, 1962
100. MILLS, C. W., *The Power Elite*, London, 1956
101. MOBERLEY, Sir W., *The Crisis in the University*, London, 1949

102. MONOD, J., 'Letter to the Editor', *Bulletin of the Atomic Scientists*, vol. 9, 1953

103. MORSE, P. M., and KIMBALL, G. E., *Methods of Operations Research*, London, 1951

104. NEEDHAM, J., and DAVIES, J. S. (eds.), *Science in Soviet Russia*, London, 1942

105. NEEDHAM, J., and PAGEL, W. (eds.), *Background to Modern Science*, Cambridge, 1938

106. NESMEYANOV, A. H., 'The Tasks of the USSR Academy of Sciences in Relation to the Fifth Five-Year Plan', *Bulletin of the Science Section: Society for Cultural Relations with the USSR*, October, 1953

107. ORD, L. C., *Secrets of Industry*, London, 1945

108. PEP (POLITICAL AND ECONOMIC PLANNING), *World Population and Resources*, London, 1955

109. PERLO, V., *Militarism and Industry: Arms Profiteering in the Missile Age*, New York, 1963

110. PIEL, G., *Science in the Cause of Man*, New York, 1961

111. PISARZHEVSKY, O., *New Paths of Soviet Science* (Soviet News), London, 1954

112. POWELL, C. F., 'International Scientific Collaboration', *World Federation of Scientific Workers Bulletin*, no. 4, London, 1955

113. POWELL, C. F., and OCCHIALINI, G. P. S., *Nuclear Physics in Photographs*, Oxford, 1947

114. PRICE, D. J. DE S., *Little Science, Big Science*, New York, 1963

115. PRICE, D. J., 'Quantitative Measures of the Development of Science', *Archives Internationales d'Histoire des Sciences*, vol. 30, 1951

116. PYKE, M., *Automation: Its Purpose and Future*, London, 1956

117. RAYLEIGH, LORD, *The Life of Sir J. J. Thomson*, Cambridge, 1942

118. READ, J., *Humour and Humanism in Chemistry*, London, 1947

119. ROSENFELD, L., 'Review: Science in History', *Centaurus*, vol. 4, 1956

120. ROTBLAT, J., *Science and World Affairs: a History of the Pugwash Conferences*, London, 1962

121. SCIENCE FOR PEACE, *Napalm* (pamphlet), London, 1952

122. 'Scientists Appeal for Abolition of War', *Bulletin of the Atomic Scientists*, vol. 11, 1955, pp. 236 f.

123. SHANNON, C. E., and WEAVER, W., *The Mathematical Theory of Communication*, Urbana, 1949

124. SHAPLEY, H. (ed.), *Source Book in Astronomy 1900–1950*, Cambridge, Mass., 1960

125. SILK, L. S., *The Research Revolution*, New York, 1960

126. SIMON, F. E., *The Neglect of Science*, London, 1951

127. SNOW, Sir C., *Science and Government*, London, 1963

128. STERNBERG, F., *The Military and Industrial Revolution of Our Time*, London, 1959

129. STEWART, G. R., *The Year of the Oath*, New York, 1950

130. STUVE, O., and ZEBERGS, V., *Astronomy of the 20th Century*, London, 1962

131. SZILARD, L., et al., 'The Facts about the Hydrogen Bomb', *Bulletin of the Atomic Scientists*, vol. 6, 1950

132. SCIENTIFIC AMERICAN, *Technology and Economic Development*, New York, 1963

133. TUGE, H., *Historical Development of Science and Technology in Japan*, Tokyo, 1961
134. UNITED NATIONS, *Economic and Social Consequences of Disarmament*, HMSO, London, 1962
135. UNITED NATIONS, *Measures for the Economic Development of Under-developed Countries*, New York, 1951
136. UNITED NATIONS, *Science and Technology for Development*, 8 vols., New York, 1963-4
137. UNITED NATIONS, *World Economic Survey, 1955*, New York, 1956
138. UREY, H. C., *The Planets*, London, 1952
139. VAUCOULEURS, G. DE, *Discovery of the Universe*, London, 1956
140. VAVILOV, S. I., *Soviet Science: Thirty Years*, Moscow, 1948
140a. VEBLEN, T., *The Theory of the Leisure Class*, New York, 1899
141. WODDIS, J., *Africa*, 3 vols., London, 1960-63
142. WALTER, W. G., *The Living Brain*, London, 1953, and Penguin Books, 1961
143. WHITEHEAD, A. N., *The Concept of Nature*, Cambridge, 1926
144. WHITTAKER, E. T., *A History of the Theories of the Ether and Electricity*, 2 vols., London, 1951-3
145. WHITTLE, Sir F., *Jet*, London, 1953
146. WIENER, N., *Cybernetics*, 2nd ed., New York, 1961
147. WIENER, N., *I am a Mathematician*, London, 1956
148. WIENER, N., *The Human Use of Human Beings*, London, 1951
149. WILSON, W., *A Hundred Years of Physics*, London, 1950
150. WOYTINSKY, W. S. and E., *World Population and Production*, New York, 1953

151. ASRATYAN, E. A., *I. P. Pavlov*, Moscow, 1953
152. AVERY, O. T., 'Studies in the chemical nature of the substance inducing transformation of pneumococcal types', *Jour. Exptl. Med., 83*, 1946
153. BALDWIN, E., *Dynamic Aspects of Biochemistry*, Cambridge, 1947
154. BANGA, I., and BALO, J., 'Elastin and Elastase', *Nature*, vol. 171, 1953
155. BERNAL, J. D., 'The Abdication of Science', *Modern Quarterly*, vol. 8, 1952
156. BERNAL, J. D., *The Physical Basis of Life*, London, 1951
157. BERNAL, J. D., 'A Speculation on Muscle', ed. J. Needham, *Perspectives in Biochemistry*, Cambridge, 1937
158. BERNAL, J. D., 'Structural Units in Cellular Physiology', *The Cell and Protoplasm*, ed. F. R. Moulton, Washington, 1940
159. BERNAL, J. D., and CARLISLE, C. H., 'Unit Cell Measurements of Wet and Dry Crystalline Turnip Yellow Mosaic Virus', *Nature*, vol. 162, 1948
160. BERNAL, J. D., and FANKUCHEN, I., 'X-ray and Crystallographic Studies of Plant Virus Preparations', *Journal of General Physiology*, vol. 25, 1941
161. BRITTAIN, R., *Let There Be Bread*, London, 1953
162. CALDER, R., *Men Against the Desert*, London, 1951
163. CLARK, F. LE GROS, and PIRIE, N. W. (eds.), *4,000 Million Mouths*, London, 1951
164. CLARKE, H. T. (ed.), *The Chemistry of Penicillin*, Princeton, 1949
165. CLEWS, J., *The Communists' New Weapon – Germ Warfare*, 1953
166. DARLINGTON, C. D., *The Facts of Life*, London, 1953
167. DARWIN, C. R., *The Effects of Cross and Self Fertilization in the Vegetable Kingdom*, London, 1876

168. DARWIN, C. R., *The Expression of the Emotions in Man and Animals*, London, 1872

169. DARWIN, C. R., *The Formation of Vegetable Mould Through the Action of Worms*, London, 1881

170. DAWES, B., *A Hundred Years of Biology*, London, 1952

171. DE CASTRO, J., *The Geography of Hunger*, London, 1952

172. DEISS, J., *The Blue Chips*, London, 1957

173. DUDLEY, Sir S. F., *Our National Ill Health Service*, London, 1953

174. DUMONT, R., *Terres vivantes*, Paris, 1961

175. DUMONT, R., *Types of Rural Economy*, trans. D. Magnin, London, 1957

176. DUTT, R. P., *The Crisis of Britain and the British Empire*, London, 1953

177. EHRENSVARD, G., *Life: Origin and Development*, London, 1962

178. FAO, UNITED NATIONS, *Yearbook of Food and Agricultural Statistics 1962*, New York, 1963

179. FISH, G., *The People's Academy*, Moscow, 1949

180. FYFE, J. L., *Lysenko Is Right*, London, 1950

181. GREEN, D. E. (ed.), *Currents in Biochemical Research*, New York, 1946

182. HALDANE, J. B. S., 'Animal Ritual and Human Language', *Diogenes*, no. 4, 1953

183. HALDANE, J. B. S., *Enzymes*, London, 1930

184. HALDANE, J. B. S., 'Genetical Effects of Radiation from Products of Nuclear Explosions', *Nature*, vol. 176, 1955

185. HALDANE, J. B. S., 'The Origin of Life', *Rationalist Annual*, 1929

186. HALDANE, J. B. S., 'On Being One's Own Rabbit', *Possible Worlds*, London, 1927

187. HALDANE, J. B. S., 'La Signalisation Animale', *Année Biologique*, vol. 30, 1964

188. HMSO, Medical Research Council, *The Hazards to Man of Nuclear and Allied Radiations*, London, 1956

189. HOPKINS, F. G., 'Analyst and the Medical Man', *Analyst*, vol. 31, 1906

190. HUXLEY, J., *Soviet Genetics and World Science*, London, 1949

191. JACKS, G. V., 'The Influence of Man on Soil Fertility', *The Advancement of Science*, vol. 12, 1956

192. KILKENNY, B. C., and HINSHELWOOD, Sir C., 'Adaptation and Mendelian Segregation in the Utilization of Galactose by Yeast', *Proc. Roy. Soc.*, vol. 139, 1951

193. LEA, D. E., *Actions of Radiations on Living Cells*, Cambridge, 1946

194. LEFF, S. and V., *From Witchcraft to World Health*, London, 1956

195. LEFF, S., *The Health of the People*, London, 1950

196. LWOFF, A., *L'Evolution physiologique*, Paris, 1943

197. MCGONIGLE, G. C. M., and KIRBY, J., *Poverty and Public Health*, London, 1936

198. MADISON, K. M., 'The Organism and its Origin', *Evolution*, vol. 7, 1953

199. MAHALANOBIS, P. C., 'National Income, Investment, and National Development'. (Summary of a lecture delivered at the National Institute of Science of India, at New Delhi, 4 October 1952)

200. MAN CONQUERS NATURE (SCR pamphlet), London, 1952

201. MICHURIN, I. V., *Selected Works*, Moscow, 1949

202. MILLER, S. L., 'A Production of Amino Acids Under Possible Primitive Earth Conditions', *Science*, vol. 117, 1953

203. MORTON, A. G., *Soviet Genetics*, London, 1951

204. MULLER, H. J., 'How Radiation Changes the Genetic Constitution', *Bulletin of the Atomic Scientists*, vol. 11, 1955

205. NATIONAL ACADEMY OF SCIENCES – NATIONAL RESEARCH COUNCIL, *The Biological Effects of Atomic Radiation*, Washington, D.C., 1956

206. NEEDHAM, J., *Biochemistry and Morphogenesis*, Cambridge, 1942

207. NEEDHAM, J., *Chemical Embryology*, 3 vols., Cambridge, 1931

208. NEEDHAM, J., *A History of Embryology*, 2nd ed., Cambridge, 1959

209. NEW BIOLOGY, no. 11, Penguin Books, 1952

210. NEW BIOLOGY, no. 12, Penguin Books, 1952

211. NEW BIOLOGY, no. 16, Penguin Books, 1954

212. OPARIN, A. I., *Life: Its Nature, Origin and Development*, Edinburgh, 1961

213. OPARIN, A. I., *The Origin of Life*, New York, 1938

214. PIRIE, N. W., 'The Efficient Use of Sunlight for Food Production', *Chemistry and Industry*, 1953

215. PRIGOGINE, I., *Étude thermodynamique des phénomènes irreversibles*, Paris, 1947

216. REPORT OF THE INTERNATIONAL SCIENTIFIC COMMISSION, *Investigation of the Facts Concerning Bacterial Warfare in Korea and China*, Peking, 1952

217. ROSEBERY, T., *Peace or Pestilence*, New York, 1949

218. ROYAL STATISTICAL SOCIETY, *Food Supplies and Population Growth*, Edinburgh, 1963

219. RÜHLE, O., *Brot für sechs Milliarden*, Leipzig, 1963

220. RUTTEN, M. G., *The Geological Aspects of the Origin of Life on Earth*, Amsterdam, 1962

221. SCIENCE FOR PEACE, 'The Export of Anti-Biotics and Sulpha Drugs to China', *Bulletin*, no. 9, 1953

222. SHERRINGTON, Sir C. S., *The Endeavours of Jean Fernel*, Cambridge, 1946

223. SIGERIST, H. E., *Civilization and Disease*, London, 1962

224. SITUATION IN BIOLOGICAL SCIENCE, THE, Moscow, 1949

225. SPURWAY, H., 'Can Wild Animals be kept in Captivity?', *New Biology*, no. 13, 1952

226. STAMP, L. D., *Our Developing World*, London, 1960

227. TINBERGEN, N., *Social Behaviour in Animals*, London, 1953

228. UNITED NATIONS, *Compendium of Social Statistics: 1963*, Statistical Papers, Series K, no. 2, New York, 1963

229. UNITED NATIONS, *The Future Growth of World Population*, Population Studies, no. 28, New York, 1958

230. VOGT, W., *The Road to Survival*, London, 1949

231. WORLD FEDERATION OF SCIENTIFIC WORKERS, *Unmeasured Hazards*, London, 1956

# Note on the Illustrations

The choice of illustrations for Professor Bernal's *Science in History* has been based on the simple principle of providing additional illumination of the text. Since the author has taken so wide a canvas on which to display his analysis, the range of illustrations has been made as broad as possible. However, science has not always been illustrated at every stage in its history and from some periods, of the few illustrations which may have existed, little or no evidence has survived to the present day; in consequence certain problems had to be solved if gaps were to be avoided. For example, virtually no original material remains of Greek science, and the scientific texts that we have are copies or translations made in later centuries. In such cases, later sources have been used if, as often happens, they make the point; Greek ideas continued for so long in western Europe that it is often still valid to use material from printed books.

In this book, where both science and the interplay of social conditions are discussed, the pictures could not always be chosen as direct illustrations of the text, but in every case it is hoped that the full captions will enable the reader to see why a picture has been chosen and appreciate its relevance, whether as allusion or analogy, by comparison or even as a comment. No attempt has been made to illustrate Professor Bernal's introductions to the various sections of his book, since this would have caused too great a mixture of subjects and historical periods. By confining illustration to the main body of the text, some degree of chronological order has been possible.

The choice of each picture has depended on a number of factors: its relevance to the text, the quality of the illustration itself, its power to provide additional visual or factual information and, of course, its aesthetic appeal. Here and there diagrams have been used, but in every case they are of historical significance. In volume I, except for the need to cast the net wide for material about Greek science, the illustrations are comparatively straightforward. Volume 2 has almost illustrated itself. Volume 3, dealing primarily with modern scientific research, is again straightforward, but volume 4 has presented some problems, in that its theme – the social sciences in history – is so wide, and that some of the concepts cannot be illustrated directly. The solution adopted has been to try, in one way or another, to complement the spirit of the text. Sources of illustrations have been given wherever possible, in a separate acknowledgements section on p. 1007.

My thanks are due to Mr Francis Aprahamian and his helpful advice, and especially to my wife, whose assistance and extensive library of illustrations has proved invaluable.

COLIN A. RONAN
*Cowlinge, Suffolk*
*June 1968*

# Acknowledgements for Illustrations

For permission to use illustrations in this volume, acknowledgement is made to the following: Aerofilms Ltd for number 252; the Air Force Cambridge Research Laboratory, Massachusetts, 244; the Atomic Energy Research Establishment, Harwell, 221; the Australian News Information Bureau, 275, 295; the British Broadcasting Corporation, 234; the Trustees of the British Museum (Natural History), 240; the British Motor Corporation, 246; the British Overseas Airways Corporation, 248; Camera Press Ltd, 299; the Cavendish Laboratory, 209, 214, 219; the Central Electricity Generating Board, 222; CERN, 262; Cheng Chen-sun, 300; Educational and Scientific Plastics Ltd and the Central Office of Information, 255; Elliott-Automation Ltd, 251; the Ford Motor Co., 245; Fox Photos Ltd, 217; Professor C. F. Foxwell, 223a; Francis Thompson Ltd, 235; Glaxo Laboratories Ltd, 269, 285; the Imperial War Museum, 257, 272; J. C. Kendrew, 278; Dr H. B. D. Kettlewell, 291; Dr A. Klug and Dr J. Finch, 281; Ku Sung-nien, 236; the Lister Institute of Preventive Medicine, 282, 283a, 283b; Liu Chih-wei, 301; Lu Hui-chun, 271; Manchester University, 212; the Marconi Company Ltd, 228, 230; Max Planck Gesellschaft, Berlin, 211; the Ministry of Defence, 243; Ken Moreman, 292; Mount Wilson and Palomar Observatories, 227; the National Institute of Oceanography, Surrey, 242; Pergamon Press Ltd, 223b; Professor D. C. Phillips, 274; N. W. Pirie, 303; Politikens Press Foto, Copenhagen, 213; Paul Popper Ltd, 290; Portland Plastics Ltd, 267; the Ronan Picture Library, 207, 208, 210, 216, 218, 220a, 220b, 226, 231, 232, 233, 249, 296; the Royal Greenwich Observatory, 215; the Royal Society, 273; Science Information Service, 225, 238, 239, 253, 256, 259, 264, 265, 276, 277, 279, 284, 287, 288, 293, 305a, 305b; the Science Museum, 229; Shell Petroleum Ltd, 254, 268, 302; the Society for Cultural Relations with the USSR, 224, 298; Time Life Inc., 294; Professor S. Tolansky, 237; Trollope & Colls Ltd, 266; Dr D. Turner, 286; UNESCO 297; the United Kingdom Atomic

Energy Authority, 260; the United States Information Service, 241, 247, 258, 261, 263; the Welsh National School of Medicine, 304; the World Health Organization, 270; Professor J. Z. Young and Dr Boycott, 289.

# Name Index

Bold figures indicate main reference

*Vol. 1: 1–364    Vol. 2: 365–694    Vol. 3: 695–1008    Vol. 4: 1009–1330*

# Subject Index

Bold figures indicate main reference

*Vol. 1: 1–364    Vol. 2: 365–694    Vol. 3: 695–1008    Vol. 4: 1009–1330*

Abacus, 120, 276, 333
Abbasids, 270, 272, 276, 349
Academia Sinica, 710
Academic freedom, 844 f., 1254, 1260 f.
Académie Royale des Sciences, 448, 450, **451**, 455, 497, 498, 517, 534, 537
Academies: in ancient world, 196 f., 207, 212 f.; eighteenth- and nineteenth-century, 514 f., 529, 549 f., 675; Renaissance, 451; seventeenth-century **450** ff.; Soviet, 709, 1190, **1265** ff. *See also* Royal Society, Académie Royale, etc.
Academy, The, **196** ff., 207, 450
Accademia del Cimento, 451 f., 455
Accademia dei Lincei, 451
Accelerator, particle, 714, 752 ff., 764 f., 843, 849 ff., 994
Advertising, 1142 f.
Aerodynamics, 811
Aeroplane, 712 f., **809–13**, 1257
Africa, 89, 140, 145, 153, 267, 269, 276, 281 f., 402, 707, 710, 723, 948, 1027, 1120, 1174, 1200 f., **1202** f., 1251 f.
Age of Reason, **531** f., 544, 1056
Agricultural Revolution, **524** f.
Agriculture: medieval, 287 f., 291 f., 312, 679; modern, 880 f., 970 ff.; nineteenth-century, 565 ff., 654 f.; origin of, 16, **91–6**, 99 ff., 342 f., 716; Roman, 224; seventeenth- and eighteenth-century, 511, **524** f.
Air-pump, 470 f.
Aix-en-Provence, 451
Akkad, 136
Albigenses, 294
Alchemy: and chemistry, 398 f., 439, 472 ff., **617** ff.; Chinese, 279 f., 1226; Islamic, 279 f.; medieval, 301, 306,

310; modern, 734 f.; origins of, 128, **222**; Renaissance, 324, 398
Alcohol, 279 f., **323** ff., 918
Alexandria, 161, 169, 209, 212 ff., 223, 227, 251, 257, 258, 270; Great Library, 231; Museum, 169, **212** f., 217, 223, 241, 272, 295, 450
Algebra, 122, 276, 332, 443
Algeria, 1120, 1130, 1180, 1199
Alpha particle, 735, 738, 752, 754 ff., 763
Alphabet, 156 f.
Alum, monopoly, 399
America: Philosophical Society, 526; War of Independence, 529, 805, 1053, 1057, 1317. *See also* United States, Latin America
American Indians, 70, 125, 535, 1032, 1241
Amplifiers, 776
Anarchism, 1108
Anatolia, 162, 342
Anatomy, 125, 188, 223, **392** f., **437**, 469, 645 f.
Animal behaviour, 71, **944** ff.
Antarctica, 799, 831
Anthrax, 565, 651
Anthropology, 1019, **1081** f., 1157, 1181 f., 1208
Antibiotics, 877, 924 ff., 929, 980
Anti-Communism, 16, 1122, 1124, 1128, 1130, 1132, 1137, 1157, 1264
Anti-ferromagnetism, 793, 854
Antioch, 251, 257, 270
Anti-particles, 764 ff., 767
Anti-Semitism, 1129, 1130
Apocalypse, 233, 402
Arabic numerals, 122, 264, 332, 1226
Arabs, 197, 209, 217, 253, 258, 266,